POLITICS AND DEVELOPMENT IN THE NORTH AMERICAN ARCTIC

POLITICS AND DEVELOPMENT IN THE NORTH AMERICAN ARCTIC

Examining the Regional Consequences of Climate Change

BY

ROMAN S. CZARNY
Jan Kochanowski University in Kielce, Poland

MAGDALENA TOMALA
Jan Kochanowski University in Kielce, Poland

AND

IWONA WROŃSKA
Jan Kochanowski University in Kielce, Poland

emerald
PUBLISHING

United Kingdom – North America – Japan – India
Malaysia – China

Emerald Publishing Limited
Howard House, Wagon Lane, Bingley BD16 1WA, UK

First edition 2021

Reprints and permissions service
Contact: permissions@emeraldinsight.com

British Library Cataloguing in Publication Data
A catalogue record for this book is available from the British Library

ISBN: 978-1-80043-717-3 (Print)
ISBN: 978-1-80043-716-6 (Online)
ISBN: 978-1-80043-718-0 (Epub)

Printed and bound by CPI Group (UK) Ltd, Croydon, CR0 4YY

ISOQAR certified Management System, awarded to Emerald for adherence to Environmental standard ISO 14001:2004.

ISOQAR REGISTERED

Certificate Number 1985
ISO 14001

INVESTOR IN PEOPLE

CONTENTS

LIST OF FIGURES AND TABLES

LIST OF ABBREVIATIONS

ACIA	Arctic Climate Impact Assessment
AEPS	Arctic Environmental Protection Strategy
AGP	Arctic Governance Project
AMAP	Arctic Monitoring and Assessment Program
ANFP	Arctic and Norther Policy Framework
CARC	Canadian Arctic Resources Committee
CAFF	Conservation of Arctic Flora and Fauna
CLCS	Commission on the Limits of the Continental Shelf
ECS	Extended Continental Shelf
IASC	International Arctic Science Committee
ICC	Innuit Circumpolar Conference
IPCC	Intergovernmental Panel on Climate Change
PAME	Protection of the Arctic Marine Environment
NWT	Northwest Territories
NASA	National Aeronautics and Space Administration
PCIJ	Permanent Court of International Justice
RCMP	Royal Canadian Mounted Police
TFSDU	Task Force on Sustainable Development and Utilization
UNCLOS	United Nations Convention on the Limits of the Continental Shelf
UNDRIP	United Nations Declaration on the Rights of Indigenous Peoples
WWF	World Wildlife Found

ABOUT THE AUTHORS

Roman S. Czarny, PhD – Assistant Professor at the Department of Law and Social Sciences of the Jan Kochanowski University in Kielce, Poland. Graduate of Jagiellonian University in Kraków (MA in English) and California State University in Dominguez Hills (MA in TESL, English Rhetoric and Composition); received his doctorate in Social Sciences at the University of Kielce; Consul of the Republic of Poland in Los Angeles (1991–1997, 2000–2004) and Chicago (1999–2000); specializes in public and cultural diplomacy, diplomatic, consular and foreign relations, international security and climate change, and the US and Arctic issues; published translator of literary and scientific books, essays, articles, film, and TV subtitles. Author of several scholarly articles and two monographs, the latest one: *The United States as an Arctic Country.*

Magdalena Tomala, PhD, hab. – Professor at the Department of Law and Social Sciences of Jan Kochanowski University in Kielce; specializes in international relations in the Baltic Sea region, the Nordic region, and the Arctic area; author of two monographs. One of them *Greenland in the System of International Relations in the Light of the Theory of World-Systems by Immanuel Wallerstein* relates to the Arctic North. Editor of two collective publications and three journals.

Iwona Wrońska, PhD, hab. – Professor of Political Science at the Department of Law and Social Sciences of the Jan Kochanowski University in Kielce. Completed her PhD and habilitation in Comparative Politics (Silesian University, Katowice); fellow at the Faculty of Social and Political Sciences, Cambridge University, UK; research associate at the Department of Political Science and EURUS, Carleton University, Ottawa, Canada; specializes in comparative dimensions of various policies comprising the North. Author of several scholarly articles and three monographs.

INTRODUCTION

It was the best of times, it was the worst of times, it was the age of wisdom, it was the age of foolishness, it was the epoch of belief, it was the epoch of incredulity, it was the season of Light, it was the season of Darkness, it was the spring of hope, it was the winter of despair.

–from Charles Dickens' A Tale of Two Cities

The quotation from Charles Dickens' book may best describe the situation of the Arctic today. On the one hand, it has become a crossroads of international politics and the center of attention for the whole world, but on the other, the region is undergoing such profound changes and at such an alarming rate that fairly soon it may be altered forever and beyond repair. It is good to be famous, but it is possibly better to be prominent for all the right reasons.

It may well be true that the Arctic is still a very mythical region, but it has recently certainly come to the forefront of world affairs. The area is going through a very important if not dramatic transformation due to climate changes which always heralded some change of direction for humanity. It is the climate indeed that is the most characteristic feature of the Arctic. The impending global warming, in a way, has brought globalization to the Arctic, which does not serve well the people living there and drastically changed the environment. The region has become famous because of the rapidly changing weather patterns. The Arctic has also served another purpose: it has brought to humanity the awareness of interconnectedness of everything that is occurring on our planet. Changing weather also made other issues very visible to the global community and especially to those in the region or close to it. Solving them seems to require a concerted effort of all interested parties, or possibly the whole world, and it remains to be seen if the damage can be mitigated or reversed at all. The problems include a plethora of concerns which will be discussed throughout this book.

Generally speaking, the new developments require a novel approach, as humanity, at least in the living memory, has not yet faced this kind of issues being equipped with pertinent knowledge. Historically, the warming period brought people to the Arctic, and the cooling era brought them demise in the region. We all seem to understand that on the one hand, the new forthcoming features provide opportunities, but on the other, they pose serious challenges. As usual, a challenge can be turned into an opportunity, while an opportunity may involve several challenges. Hence, on the principle of dualism, these two notions can be alike at any given time and simply exhibit two different aspects of the same issue, depending on the person who perceives and evaluates them. One could say that they can be agathokakological in nature, i.e. composed of both good and evil. As Baruch or Benedict de Spinoza said,

> *One and the same thing can at the same time be good, bad, and indifferent, e.g., music is good to the melancholy, bad to those who mourn, and neither good nor bad to the deaf.*
>
> *(Spinoza, 1677)*

It almost goes without saying that the topic of the Arctic is very current, if not pressing and imperative, as evidenced by the growing interest in the region and the number of publications devoted to it. It seems to be further complicated by a multitude of unpredictable determinants. The Arctic is undoubtedly very complex and multifaceted. Since we have hardly a way of knowing what the future holds, the case of the Arctic has been flooded with a growing body of commentaries, resolutions, predictions, calls to arms, and above all – the promises of unspeakable riches for some.

The state of relations between countries in the Arctic is the result of conflicts, rivalry, or cooperation between them. In the literature of international relations, we can find studies analyzing the course of these phenomena in the Arctic. On the one hand, we can point to works that address issues of cooperation in the Arctic. For example, Y. Hirose emphasizes the importance and need for ecological cooperation, despite the threats in the region, resulting from the possibility of conflict over the region's natural resources, territory, the rights of Indigenous peoples, and international business (2018, p. 38). A similar position is presented by D. Scrivener, who in his article focuses on the period from 1996 to 1998 – as a transition phase in the process of circumpolar Arctic environmental cooperation. He stresses the importance of the eight States forming the Arctic Council (Scrivener, 1999). A. Østhagen, on the other hand, emphasizes cooperation across maritime borders despite the discrepancy between the aspirations and reality of pan-Arctic cooperation (2020). An interesting approach to an Arctic cooperation is presented by O. Young in a document prepared in connection with the Fourth Conference of Parliamentarians of the Arctic Region in Rovaniemi from August 27 to 29, 2000. It points out the key issues for an Arctic cooperation, such as the institutionalization of cooperation on the model of the Antarctic, effective governance in the Arctic, and cooperation with international organizations. In addition, O. Young presents recommendations for the Arctic Council to strengthen and consolidate the activities of this organization (2000). John Higginbotham of the Centre for International Governance Innovation suggests that in order to improve the economic development, the North American Arctic States should start forming alliances. By forming a united front, they will be stronger in attracting private sector (Weber, 2018). On the other hand, research on rivalry and conflict in the Arctic occupies an important place in the literature. Elena Wilson Rowe analyzes this sphere, focusing on the perception of the Arctic in the press media. She draws attention to popular headlines in the press, such as "The Race for the Arctic," "The New Cold War," which show this territory as a source of potential conflict (2013). As K. Keil notes, gas and oil play a key role in relations between the five Arctic States, and gaining access to them implies confrontation between Arctic stakeholders (2014, pp. 162–190). In turn, Mikkel Runge Olesen points out that the only possible conflict may emerge between Russia and the West. At the same time, he also recognizes that China, the most influential non-Arctic player, does not currently have sufficient reasons to initiate conflict on a military level (2017, p. 3). It is also interesting to see the Arctic in the literature as a space of competition and an arms race. For example, L. E. Flake highlights the role of Russia as the only non–North Atlantic Treaty Organization (NATO) Arctic riparian state, which undoubtedly has the greatest influence on the possibility of disputes in the region. He draws attention to two of the most serious issues of conflict, namely: the designation of the seabed and the right to navigation (2015).

These publications show the complexity of relations in the Arctic region. Many authors note this dichotomy, usually pointing to the opportunities and dangers arising from mutual relations between state and nonstate actors involved in Arctic affairs. For example, O. Young in his book *Arctic Politics: Conflict and Cooperation in the Circumpolar North* stresses the extreme importance of the region not only for natural scientists but also for the social sciences because of the complex relationships both in the subjective and objective sense (1992, p. X). I. G. Brosnan, on the other hand, has undertaken in his work to study the possibilities of cooperation in the Arctic on the basis of strategic declarations of five Arctic States. He identifies areas that should foster the development of cooperation, such as scientific research, Indigenous peoples' sovereignty issues, environmental issues, resource development, and maritime transport. Despite these incentives and benefits of Arctic cooperation, he stresses that Arctic conflict is by no means inevitable (Axelrod et al., 2011). According to O. Young, the biophysical changes taking place in the Arctic are a premise for reassessing the effectiveness of existing arrangements and asking for new forms of Arctic governance. In his article, the author advocates an effective tripartite cooperation set of solutions including an agreement to repeal without expiration, the claims for extended jurisdiction of the continental shelf by littoral states, adaptation of the Arctic Council to meet emerging challenges in the Arctic, and a desire to develop regulatory systems to address the problems associated with shipping, fishing, and offshore oil and gas extraction (2009). The dilemma between cooperation and rivalry is addressed by T. L. Palosaari in the article "The Amazing Race: On resources, conflict, and cooperation in the Arctic." The author shows the factors determining the peaceful development of cooperation in the Arctic which contrast with the Arctic race for resources in the region. He compares the significance and weight of Arctic problems, emphasizing that the international community should focus on solving human and environmental security problems instead of addressing the issue of borders or military capabilities (2012).

An interesting approach is presented by M. Łuszczuk, who in his work *The Evolution of International Roles in the Arctic* indicates the stages of development of the Arctic as an environment of international roles. He proposes to explain the periodization of the development of the Arctic in the nineteenth and twenty-first centuries as a regional subsystem of international relations, identifying four stages of the development of the Arctic: the Arctic as a territory of conquest, the Arctic as a space of competition, the Arctic as a sphere of cooperation, and the Arctic as an arena of challenges (Łuszczuk, 2015). The above research shows that international relations in the Arctic can and should be analyzed, taking into account the consecutive stages and current structure, which lead from conquest, through cooperation, to rivalry in the region. Thus, game theories can explain the relations between the subjects (players).

As usual, what is so promising for one may spell disaster for the other. The warming Arctic does not bode well for the traditional way of life of the Natives; the exploration and exploitation of the natural resources together with intensifying maritime traffic on the once pristine seas cannot be good for the environment, and the climate changes seem to be threatening the status quo of the whole planet. In essence, opportunities and challenges are really in the eye of the beholder, to paraphrase the famous saying, and only time will eventually evaluate which category is going to come on top, and for whom. As with all things unknown, there has been much speculation of the possible outcomes of the dynamically developing situation.

The Arctic is a space where one can observe indications of cooperation and conflict. The question can be asked, which option will win? Palosaari et al.

indicate that it is more beneficial for the Arctic States to focus on the environment and population of the region, but Olsen or Flake notice the factors which clearly herald conflict between the players. In this context, the key approach used in this book will be J. Nash's equilibrium theory.

Game theory is one of the basic tools for scientific analysis of social phenomena. John Nash's "prisoner's dilemma" consists of a two-player nonzero sum game in which each player can gain by betraying his opponent, but both will lose if they do betray (see Table 1).

It is worth noting that in this concept the conflict strategy prevails over the peace strategy. Within the prisoner's dilemma, there are four options:

(1) Cooperation/cooperation (−1m−1m = −2m)

(2) Cooperation/conflict (0m − 10m = −10m)

(3) Conflict/cooperation (−10m + 0m = −10m)

(4) Conflict/conflict (−5m−5m = −10m)

A rational approach, based on obtaining the greatest possible benefits, tells players that they should betray. Only one option out of four provides for Nash's equilibrium. It is a situation in which both players cooperate with each other. An important element that should be emphasized in the context of relations between States is that as a result of cooperation both gain. As M. Wajzer points out,

> *In political science it is used in studies of electoral and legislative behavior, in analyses of processes of forming political coalitions and in examining issues related to democratization, national security and armed conflicts.*
>
> *(2018, pp. 61–70)*

Therefore, this approach will be used in the work in two ways.

Firstly, the theoretical approach will allow for the analysis of Arctic partners' behaviors in subsequent rounds of the "Arctic game." The approach of

Table 1. The Payout Matrix in the Prisoner's Dilemma Concept.

	B – conflict	B – cooperation
A – conflict	−5; −5	−10; 0
A – cooperation	0; −10	−1; −1

Robert Axelrod will be used in this aspect, who invited academics from all over the world to participate in the tournament for computer programs playing the prisoner's dilemma. The obtained programs varied in terms of complexity, initial behavior, reaction to the opponent's action, etc. The results showed that in multiple games, selfish strategies provided on the average very small winnings compared to the more altruistic ones. This is due to the fact that it pays off for the States to work together to find a compromise which satisfies the conflicting parties. On the other hand, a strategy in which one of the players betrays is profitable, but only short-term. One player who deceives his partner in the first round gains an advantage over the other. However, as simulations carried out by R. Axelrod showed, selfish games gave very poor results in subsequent rounds. It will, therefore, be important to show the relations between the Arctic States over time, i.e. since the early 1990s, when we can talk about a new stage of relations between the States in the Arctic not only in the context of their our own coalition but also relations with other entities (Axelrod et al., 2011).

Secondly, when analyzing the literature on the subject, one should pay attention to the issues of coalition and division axes in the Arctic. The two-block system, which prevailed in this region since World War II (WWII) until the fall of the Berlin Wall, divided the region into the American and Soviet zones of influence. The symbolic fall of the "Iron Curtain" changed the situation in the region. First of all, we are now dealing with a multilateral system, which allows for various possibilities of cooperation, rivalry, and conflict between state and nonstate actors. Second, multilateralism encourages the establishment of a coalition; hence the various manifestations of the coalition, such as the Arctic Five, the Arctic Eight, non-Arctic States, and others, can be identified in literature. The aim of the analysis will be, first of all, to examine the coalition capabilities of the three North American Arctic entities in terms of cooperation in the field of environment, sustainable development, Indigenous peoples, and others.

A comparative method of analysis will be useful to identify a space of common or divergent interests of the United States, Canada, and Denmark.

As a result of the analysis, it will be possible to point to one of the axioms indicated by John F. Nash, which could be a good arbitration scheme:

(1) The axiom of rationality – the solution should be found in the negotiation set;

(2) The axiom of independence – the States hold on to their goals;

(3) Axiom of symmetry – the symmetrical problem of negotiation should determine the result which does not distinguish between the parties concluding the agreement; this axiom reflects the equal potential and negotiation skills of the players;

(4) Axiom of independence from unrelated alternatives.

The question can be asked: does rivalry outweigh cooperation in the North American Arctic or is it the other way round? do the entities manage to gain the benefits of cooperation?

One can pose the hypothesis that the States and Arctic actors should cooperate with each other in the light of the prisoner's dilemma of obtaining tangible benefits, but the more probable, and definitely more possible variant of no such cooperation or breaking the cooperation is rivalry, which in the short term gives an advantage over other players, but in the long term causes losses.

The analysis of the behavior of international relations entities in the Arctic will allow comparing the areas of cooperation and rivalry. It will, thus, make it possible to identify the best development strategy for the region.

In our work, we should like to adopt an interdisciplinary approach. As Barry Buzan and Mathias Albert indicate,

> *International Relations occupies the middle ground, being comprised of two rather vague terms: international, which at least hints at actors other than the state; and relations which opens the possibility of more than just politics.*
>
> *(2017, p. 4)*

We would like to emphasize that the nature of the study concentrates on international relations and covers political, economic, and societal aspects of the main players. Thus, it makes an opening for international relations as a multidisciplinary construction through involving state and nonstate actors, including the specific Indigenous population, which is important for building international cooperation in the region.

The structure of the book is based on the parameters of the game theory. Hence, Chapter 1 outlines the arena of the game: The North American Arctic. It starts with describing the geographic features of the Arctic and explains the effects of climate change as a predominant factor that influences growing international attention to the region. It also explains the status of the Arctic within the frame of international law. The North American Arctic experiences similar effects of climate change as the whole Arctic but has distinctive political, social, and economic features which might determine cooperation or competition.

Chapter 2 captures the leading subnational and national players or decision-making bodies in the North American Arctic and analyzes their powers and policies. The analysis proceeds in two stages. The first stage focuses on subnational governance in Canada, Denmark, and the United States. It attempts to present the value of devolution of power that allocated more responsibilities to subnational governments, thus significantly transforming the governance in the North American Arctic in the last 50 years. The second stage of the analysis pays attention to national strategies and policies developed by the North American Arctic States in the last decade. The two intertwined Canadian strategies of 2009 and 2019, Denmark's 10-years strategy of 2011, and the first presidential strategy of the Barack Obama administration launched in 2013 will be primarily examined.

Chapter 3 indicates in which areas countries are seeking and giving up cooperation. It should also be stressed that the analysis will cover the period from 1990 to 2020, i.e. a period of 30 years in which we will examine how entities behave when one of the parties refuses to cooperate and whether the others still want to cooperate or follow the first entity. Such an approach will allow us to estimate to what extent we are dealing with a stable region and indicate its development possibilities. In addition, it will also allow us to examine whether the same processes are taking place at international level as in the case of individual decisions.

To sum up, it can be emphasized that the study will allow for both subjective (selection of areas of cooperation and competition in the Arctic done by the authors) and objective analysis. The latter will make it possible to identify those entities that seek to achieve their own benefits or those that aim to cooperate across borders.

Although the book draws from many sources, some established scholarly, and some less so, and employs many opinions through quotations – the syntheses, conclusions, and general statements are authors' own, for which we take full responsibility. The authors are aware that this work is far from exhausting the subject because tomorrow may bring a set of completely new issues; neither is it a blog, especially in view of the everyday dynamic developments in the Arctic. Some specialists may find the book lacking in depth, be it experts in international law, meteorologists, or military historians, but the work is addressed to the general audience, educated but not necessarily well-informed, and equipped with notes for students and enthusiasts of the subject.

As Jerome K. Jerome (Jerome, 2018, p. 28) said in 1946: "Dear Reader, You and I are each the center of the universe in our respective opinions. You, as I understand it, were brought into being by a considerate Providence in order that you might read and pay me for what I write; while I, in your opinion, am an article sent into the world to write something for you to read." We fully subscribe to this point of view, with the notable exception of hoping to be paid for what we write – we would be thankful just for the Readership.

1

SETTING UP THE STAGE: THE NORTH AMERICAN ARCTIC

INTRODUCTION

The Arctic has been very much present in the relatively recent international politics. Although initially perceived as an area of the Cold War competition, after the geopolitical transformation and climate change it became a subject of international competition and cooperation. Chapter 1 deals with geography, history, and perceptions of the Arctic and examines definitions as well as parameters and the notion of sovereignty pertaining to the region, not exuding the all-important issue of economy. It attempts to present the terminology as accurately as possible and as necessary for the purpose of this work, and consists to a large degree of a review of the literature, with some preferences of the authors. It is absolutely necessary to establish the parameters, or the metalanguage, in order to discuss the issues connected with the Arctic, with the absolute certainty and full understanding that even then, there will remain doubts and diverging opinions. This section attempts to present what is understood by the term "the Arctic" and especially what constitutes the North American Arctic today. By this it delineates the stage for the application of the Nash game theory.

UNDERSTANDING THE ARCTIC

Geography: What Exactly Is Arcticness and the Arctic Itself?

If "Arcticness" indeed is a term, as coined in the book titled *Arcticness: Power and Voice from the North,* edited by I. Kelman and published by UCL Press in August 2017, then it must relate to examining and exploring opportunities and limitations in engaging with the Arctic under change, and the Arctic peoples experiencing the change.

> *As the world is increasingly looking northwards to a region undergoing rapid change, identifying what, who or where has the 'quality of being Arctic' is high on the agenda; for actors from both near and far, their 'Arcticness' becomes a potential asset as they position themselves for Arctic futures.*
>
> *(Medby, 2017)*

Therefore, Arcticness denotes all things Arctic, which is very useful as it does not require any further explanation. Being undefinable, it is open to interpretation for and by anyone.

Since the age of the classical Greek philosophers, the Arctic has held a mysterious fascination for Western civilizations, which often portrayed this mythical region with fanciful images that belied reality. The word "Arctic" itself originated from the Greek word *arktos* meaning "bear," and was used to describe the land and seas lying to the north under the constellation Ursa Major, the Great Bear (Grant, 2010, p. 5). It is also related to the constellations of Little Dipper (or Little Bear) and Big Dipper (or Big Bear), which are located near the Polaris – North Star (Makowski & Rossa, 2011, p. 195).

The term "The Arctic" means "a lot" and then more. The literature is far from consistent in delineation of the borders and territorial reach of the areas under the common term.[1] Many scientists have written on the subject; therefore, this chapter consists mostly of a review of the literature, with some preferences of the author.

Even members of the Arctic Governance Project (AGP) in the part "Defining the Arctic" state the following[2]:

> *There is no universally accepted definition of the Arctic. We follow the practice of the Arctic Council in treating the Arctic as a*

---

1 "Definitions," "As Europeans Understand It" and, partially, "As North Americans Understand It" are based on Ryszard M. Czarny's book titled *The Imperative High North: Opportunities and Challenges*, Michal Vaško – Vydavatel'stvo 2013, Chapter 1. The High North, pp. 11–22. In addition, some portions, where noted, come from *The High North: Between Geography and Politics*, Springer International Publishing Switzerland 2015 by the same author (Czarny, 2013).

2 The Arctic Governance Project (AGP) is an unofficial initiative supported by a group of private funders and intended to bring together preeminent researchers, members of the policy community, and representatives of Indigenous Peoples in the interests of exploring ways to achieve a sustainable and just future for the Arctic. The term "we" in this report refers to the members of the AGP's Steering Committee and the Executive Secretary.

> *circumpolar region encompassing both marine and terrestrial systems extending southward from the North Pole, covering about 8 percent of the Earth's surface, including areas located within the jurisdiction of eight States, providing a homeland for many indigenous peoples, and including altogether some 4 million residents. But this region is highly diverse in biophysical, socioeconomic, and cultural terms.*
>
> *(Arctic Governance, 2010)*

The Arctic is said to have suffered most because of the changes in weather conditions and, undoubtedly, due to human activities contributing to the climate transformation. In the vein of complaints about the weather, people usually attribute extremely cold flurries to that region and blame the Artic for bringing it to them.

> Arctic *is the northernmost region of the Earth, centred on the North Pole and characterized by distinctively polar conditions of climate, plant and animal life, and other physical features. It has sometimes been used to designate the area within the Arctic Circle a mathematical line that is drawn at latitude 66°30′ N. marking the southern limit of the zone in which there is at least one annual period of 24 hours during which the sun does not set and one during which it does not rise. A generally useful guide is the irregular line marking the northernmost limit of the stands of trees.*
>
> *(Encyclopedia Britannica, 2020)*

As for the definition of Arctic States, the situation appears to be slightly more complicated. After all, Sweden and Finland are noncoastal Arctic States, and so is Iceland, otherwise called nonlittoral states, but they still are all permanent members of the Arctic Council.[3] Therefore, depending on the context, one talks about the Arctic Five (Canada, Denmark/Greenland, Norway, Russia, and the United States) or the Arctic Eight (the Arctic Five plus Finland, Iceland, and Sweden).

The Arctic is a part of our globe encompassing both lands and waters, and including the polar and subpolar zones in the Northern Hemisphere around the North Pole. It is one of the wildest and at the same time least explored and accessible places on Earth. The names of its seas and rivers are not widely known, although the Siberian Yenisei and Lena Rivers are not only huge, but they also carry more water to the sea than, for example, the Mississippi or the

3 For more on the Arctic Council, see Chapter 3 and http://www.arctic-council.org/.

Nile. Greenland, the biggest island in the world, is six times bigger than, for example, Germany, but it is populated by only 57,000 people, mostly by the Inuit inhabiting small coastal settlements. In all the Arctic, also called the Arctic Circle (with a small margin in the South), nearly half of the population of 4 million lives in a few post-Soviet cities like Murmansk or Magadan. This region, to a large degree untouched, L. Sykulski (2009, pp. 6–7) defines as "an area of the globe around the North Pole. There exist no clearly defined borders of the Arctic." Its so-called astronomical border is the Arctic Circle (66°30′40″ N) covering the area of approximately 21 million square kilometers, in which there occurs a phenomenon of the polar night, where the northern lights can shine both "day" and night.[4]

The Arctic territory consists of seas, islands [among others, Greenland, Baffin Island, Novaya Zemlya, Victoria Island, Ellesmere Island, or Svalbard Archipelago] – together with the tundra vegetation covered stripe of Eurasia, the northern and eastern part of Labrador, the land part of northern Canada, and Alaska, straits and bays, and it occupies some 27 million km^2. The land part, encompassing the tips of the East European Plain, the West European Plain, the West Siberian Lowland, Kolyma Lowland, Yano-Indigirskaya Lowland, as well as the plains at the Arctic Ocean in North America, occupies some 10 million km^2. Mt Gunnbjörn [3,700 m above sea level] in Greenland is considered the highest peak of the Arctic.

The Arctic is characterized by lower average air temperatures of a very wide range – annual average temperature in Reykjavik is listed at 4°C, in the middle of the Arctic Ocean −18°C, and in the upper parts of Greenland continental glacier – at −29.1°C (ACIA, 2005, p. 10) and lower water temperatures, as well as the presence of ice and snow sheet – seasonal at lower latitudes, and permanent at higher ones (Węsławski, Stempniewicz, & Kwaśniewski, 2007, p. 312).

As a consequence of the harsh or even extremely cold climate, the flora of the Arctic is rather poor, dominated by

> *...dwarf shrubs, hardy perennials [mostly grasses and sedges], moss, lichen, and moss-tundra, dwarf-shrub-tundra, lichen-tundra, and forest-tundra. The animal world is represented mainly by the abundant though little diversified marine fauna consisting mostly of fish. Although there exist here some 150 fish species, because of their numbers alone, the fish play important role in the ecosystem functioning, as well as the economies of the Arctic States.*
>
> *(Czarny, 2015, p. 14)*

4 It is such a broad interpretation that it lacks common acceptance.

The area under examination is also of great strategic importance due to a great abundance of natural deposits, among others, of crude oil and natural gas.

The High North, which is sometimes referred to as Far North,[5] is one of the key areas on a global scale whose importance cannot possibly be overestimated, although the knowledge of an average American, Canadian, or European on the subject might be somewhat limited. In the era of high prices of energy resources and climate warming, the region has experienced a high level of activity and also a certain nervousness of the States of that region, both of which influence international relations on a scale going beyond the interests of one hemisphere only. In practice, that involves a clear orientation of interests, and not political rhetoric only, the more so as the region truly possesses a high potential allowing for economic growth in several sectors. The region boasts of a richness of the fauna – very large fishing stocks, which drives a dynamic development of the fishing industry. A constantly growing demand in the international market for the high-quality white fish from the Barents Sea made fishing these waters a multibillion dollar industry (www.norden.org, p. 2).

As Europeans Understand It

There exist several other definitions and ways of denoting the Arctic which are based on a variety of factors such as average temperature, the northern tree line, the extent of permafrost on land, the extent of sea ice on the ocean, or jurisdictional or administrative boundaries (Hassol, 2004, p. 4; Yong & Einarsson, 2004, pp. 17–18).

Some researchers are of the opinion that the astronomical borderline is one way of delineating the Arctic, but there exist more ways of defining the geographical scope of the region:

- The climatological border running along the +10°C isotherm on land and +5°C on sea in the warmest month of the year (July),[6]

5 The term "Far North" is far from clear as it may refer to the following: Far North (Russia), a part of Russia which lies beyond the Arctic Circle; Far North Alaska, United States; Far North (Canada); Far North, Chile, one of the five natural regions of Chile according to CORFO; Far North District, the administrative district in northern Northland Region, New Zealand; Far North Region (Cameroon), Cameroon; Far North (South Australia), a region; Far North Queensland, Australia; Far North (New Zealand electorate), a former New Zealand Parliamentary electorate; Vhembe District Municipality in South Africa, also known as Far North.

6 In this case, the Arctic encompasses ca. 26.5 M km^2.

- The so-called Nordenskiöld Line,[7] also termed a botanical border, delineated on land by the northern tree line (coinciding to a large degree with the climatological border); and in water environment demarcated by water temperature and reach of glaciation,
- The geopolitical border which, according to Osica (2010, p. 12) "concerns a region extending north of the 60th parallel. As well as this, there is the notion of the Arctic Circle, which narrows down the subregion's political range to the Polar Circle. From this perspective, the major players are those States of the Arctic G5 which neighbor with the Arctic Ocean, i.e. Russia, the USA, Canada, Denmark/Greenland and Norway." O. Osica (2010, p. 12), presenting the American perception of the issues, adds the following: "When viewed from the perspective of the term 'High North,' the circle of concerned parties is increased with the Polar Circle States of the Arctic Council, whose territories are adjacent to the Arctic circle or are located below it, namely Iceland, Finland and Sweden."

As North Americans Understand It

The definition presented by American researchers in the Congressional Report of 2018 was formulated by Ronald O'Rourke:

> *Some observers use the term 'high north' as a way of referring to the Arctic. Some observers make a distinction between the 'high Arctic' – meaning, in general, the colder portions of the Arctic that are closer to the North Pole – and other areas of the Arctic that are generally less cold and further away from the North Pole, which are sometimes described as the low Arctic or the subarctic.*
>
> *(O'Rourke, 2018, p. 4)*

The American point of view on the matter emphasizes that many definitions of the Arctic spring from various descriptions of the land and maritime areas encompassed by this region. It is also used in a variety of ways in political discourses in which the employed terms may carry different meanings. For example, the CRS Report (O'Rourke, 2012), although not based on a single definition, still attempts to provide one:

7 Nordenskiöld, Adolf Erik (1832–1901), baron, Swedish researcher and explorer who, among others, led the Arctic expeditions on Spitsbergen in the years 1864–1873.

> *The most common and basic definition of the Arctic defines the region as the land and sea area north of the Arctic Circle (a circle of latitude at about 66.34° North). For surface locations within this zone, the sun is generally above the horizon for 24 continuous hours at least once per year (at the summer solstice) and below the horizon for 24 continuous hours at least once per year (at the winter solstice). The Arctic Circle definition includes the northernmost third or so of Alaska, as well as the Chukchi Sea, which separates that part of Alaska from Russia, and U.S. territorial and Exclusive Economic Zone (EEZ) waters north of Alaska. It does not include the lower two-thirds or so of Alaska or the Bering Sea, which separates that lower part of the state from Russia.*
>
> *(O'Rourke, 2018, p. 5)*

In turn, the definition adapted by the Arctic Monitoring and Assessment Programme (AMAP) states that the Arctic[8]

> *...essentially includes the terrestrial and marine areas north of the Arctic Circle (66°32′ N), and north of 62° N in Asia and 60° N in North America, modified to include the marine areas north of the Aleutian chain, Hudson Bay, and parts of the North Atlantic, including the Labrador Sea.*
>
> *(AMAP, 2020)*

In North America, the term "Arctic" omits vast areas of treeless tundra located down south of the Arctic Circle, although it has the physical attributes of a polar landscape.

> *Similar discrepancies arise when this definition is applied to Europe and Asia, where one finds forests and warmer climates north of the Arctic Circle. Some scientists prefer to define the Arctic as the area north of 10°C July isotherm, or in lay terms, north of the region that has a mean July temperature of 10° Centigrade. Geographers tend to favour boundaries set by the presence of continuous permafrost, as opposed to the discontinuous permafrost found in the Subarctic. If global warming continues, the boundaries set out in the last two definitions will change. Although also likely to be affected by climate change, the treeline definition is more appropriate to*

8 AMAP is a work group of the Arctic Council. If one clicks on "About AMAP," and then goes to the tab "Geographical Coverage," one may follow the discussion on the subject.

> *North America as it marks the boundary between Inuit traditional homelands and those occupied by northern Indians.*
>
> *(Bone, 1992, pp. 17–27)*

In view of the above, one can differentiate three different Arctics: European Arctic (High North, *Norden,* Northern Territories), Russian Arctic (Russian Far East – RFE), and North American Arctic (NAA), the latter composed of Alaska, Canadian Arctic areas, and Greenland which, geographically speaking, is located in North America.

John Higginbotham and Jennifer Spence, in their Special Report titled *The North American Arctic: Energizing Regional Collaboration and Governance,* point out to the differences between the three areas listed above due to some fundamental geophysical and political differences among the Nordic, Russian, and North American subregions of the Arctic, which can limit the focus and effectiveness of cooperation in important policy areas.

> *For example, the Nordic countries (Denmark, Finland, Iceland, Norway and Sweden) benefit from open water all year round, strong transport and energy infrastructure, well-developed industries and cities, and well-established and long-standing regional cooperation among themselves and with Russia. Meanwhile, the Russian Arctic has had strong, highly centralized Arctic development policies and has invested heavily in cities, industries, surface and marine transport, resource development and military capabilities. Russia is working hard to establish its hegemony over the commercialization of the Arctic Ocean. In contrast, the NAA region – Alaska, Yukon, the Northwest Territories, Nunavut and Greenland – has harsher and more extreme environmental conditions, spotty or non-existent infrastructure, and is composed primarily of small, remote communities with economies that have suffered from boom-and-bust development. Furthermore, the American and Canadian Arctic regions are being pushed and pulled in opposite development directions by changing federal priorities.*
>
> *(Higginbotham & Spence, 2018, p. 2)*

In the North American Arctic, with one exception, the Arctic lands presently claimed by the United States, Canada, and Denmark are considered secure. Still outstanding is the dispute over ownership of Hans Island in Kennedy Channel, lying midway between Ellesmere Island and Greenland. Hans Island is only 1.3 square kilometers in size and was once considered a large pile of barren rocks. Other disputes relate to foreign rights of transit in adjacent waters, particularly the Northwest Passage, the offshore boundary in the Beaufort Sea, submarine activity under the sea ice, and rights to seabed a nation's territorial waters and continental shelf.

The Issue of Sovereignty of the Arctic

> *A major intellectual requirement of our time is to rethink the question of sovereignty – not to weaken its essence, which is crucial to international security and cooperation, but to recognize that it may take more than one form and perform more than one function.*
>
> *–Former Secretary-General of the United Nations, Boutros Boutros Ghali*

As sovereignty has so many definitions which vary to a considerable degree, the authors have chosen to select the one presented by the United Nations.

> *State sovereignty, in its most basic sense, is being redefined – not least by the forces of globalisation and international co-operation. States are now widely understood to be instruments at the service of their peoples, and not vice versa. At the same time individual sovereignty – by which I mean the fundamental freedom of each individual, enshrined in the charter of the UN and subsequent international treaties – has been enhanced by a renewed and spreading consciousness of individual rights. When we read the charter today, we are more than ever conscious that its aim is to protect individual human beings, not to protect those who abuse them.*
>
> *(S-G UN, 1999)*

> *Sovereignty in and of itself is not juxtaposed to international law or multilateral cooperation. In fact the basic idea of sovereignty is central to the founding principles of the UN. The UN serves as an intergovernmental organization where independent nations come together to work in cooperation for their national interests. The framers of the UN did not intend to create a world governing body but rather a body where independent nations could gather to discuss issues in an increasingly globalized world. In this environment, a*

> *number of UN Member States have increased efforts to insert language around sovereignty within UN decisions, otherwise known as Resolutions, and in particular those dealing with human rights issues (...) In the current attempts to privilege 'sovereignty' in multilateral negotiations on human rights issues, Member States in reality are pushing for the UN to ignore national human rights violations. These States want to continue to discriminate against marginalized communities with impunity under the guise of "national interests" and "cultural values".*
>
> *(OutRight, 2017)*

The discussion over the possession of Arctic resources and authority over the sea routes brought back the issue of sovereignty over the North. As I. Head indicates, territorial sovereignty is a compound of several factors, involving both rights and responsibilities. The sovereign has the right to control access to the territory and exercise its authority, and bears the obligation of protecting it. The exercise of these State activities not only accrues lawfully but the manifestation of them is indicative of who is the sovereign. Just a claim to territorial sovereignty if not accompanied by such a manifestation might be challenged by other States (Head, 1963, p. 224). These disputes have a long history but now they have emerged again.

Sovereignty in the Arctic has long ago transcended the Westphalian notion of States and their properties. After all, contrary to some myths and simple lack of knowledge, it is not land per se but a combination of terra firma and lots of water, mostly in the form of ice. And on top of everything else, territorial disputes may well disregard what is on top – they are rather about what is on the very bottom, including the sea.

With all the talk about sovereignty, one cannot forget that this is about people, inheritance, tradition, discoveries, claims made historically to the portions of a specific territory, and their occupation (in a good sense, if that is at all possible). There is an ugly word for territorial expansion which is "colonization." Centuries ago it used to be much commendable and lavishly rewarded by the rulers, while they themselves received incredible riches and power.

> *The history of Arctic sovereignty in North America belongs to more than just the Inuit, Britain, Canada, the United States and Denmark. For over 300 years, it also involved the Dutch, Russians, Portuguese, French and Spanish Basques, all vying for control over the Arctic seas of the Old and New Worlds. Success or*

> *failure depended on a number of factors: the effect of European wars and internal conflicts, adoption of new technologies, accessibility of resources and market demand; the Protestant Reformation; as well as changes in economic and political power within the global community. Overriding all else during the late seventeenth and eighteenth centuries was an event beyond anyone's control – the cumulative effect of the Little Ice Age – when cooler temperatures increased the area and depth of Arctic sea ice, temporarily halting exploration. By the time a warming trend reopened the northern waters of Davis Strait, shifts in economic and military power had changed the world map. Not only were different nations competing for control of the Arctic, but their objectives and priorities had also changed.*
>
> *(Grant, 2010, p. 56)*

One significant concept in discussion of the history of Arctic sovereignty is the *terra nullius* principle, meaning uninhabited or "nobody's lands." At the time of discovery, Europeans referred to the lands of the New World as *terra nullius* in spite of sizable aboriginal populations, based on the argument that indigenous peoples were uncivilized since they were pagans and without a structured form of government, thus could not be considered rightful owners of their lands (Berger, 1991; Grant, 2010, p. 14). At no time was any thought given by States to secure the consent of indigenous peoples to legitimatize discovery claims. Moreover, force was considered acceptable if the natives resisted subjugation. As explained by Olive Dickason, proprietary rights were set out in the laws of an organized State, whereas "the land of pre-state people without such law was therefore legally vacant," or *vacuum domicilium* (Dickason, 1989, pp. 234–236). The requirement of a permanent settlement to protect a discovery claim did not appear in a written document until the 1884–85 Berlin Conference on Africa, which introduced a Principle of Effectivity and created the rules of effective occupation of the conquered lands (General Act, 1885). It means that a State could claim colonies if they effectively possessed them by signing treaties with the local population and establishing administration. Otherwise, such a claim would be "inchoate" or temporary, and thus vulnerable to challenges by other nations. This rule provided the basis for future interpretations of Arctic sovereignty. Danes confirmed the effective occupation by supporting the settlement but while Russians and the British by organizing fur trade stations, taken over by Americans and Canadians in the end of the nineteenth century.

In 1905, the renowned English authority Lassa Francis Lawrence Oppenheim further confirmed the "effective occupation principle" by declaring in his Treatise that a territory may be acquired through occupation "if inhabited by natives whose community is not considered as a State" even though "natives may live on a territory under a tribal organization." He argued that any agreement with indigenous peoples to cede their perceived rights "is usually neither understood nor appreciated by them" and that even if they did understand it, "such agreements had only 'a moral value'" (Oppenheim, 1905–06, p. 275). A similar approach can be found in William Edward Hall's writings (Hall, 1894). As Grant summarizes, both Oppenheim and Hall argued that discovery claims alone were not sufficient to maintain title to uninhabited lands and must be followed within reasonable time by some form of effective occupation. But Oppenheim went further, arguing that this could be accomplished only by actual settlement accompanied by administrative acts. Otherwise acts of possession by themselves constituted "fictitious occupation only, unless there is left on the territory a settlement which is able to keep up the authority of the flag" (Grant, 2010, p. 19).

Consequently, the Canadian Inuit were never asked to sign a treaty and were never represented under the Indian Act, in spite of a Supreme Court decision in April 1939 that "Eskimos were Indians" and thus should be considered wards of the federal government (Backhouse, 1999; Diubaldo, 1981).

To strengthen its right to unexplored lands Canada formulated an Arctic sector principle or sector theory that was based on a meridian of longitude.[9] The theory was formulated by Canadian Senator Pascal Poirier and presented in the Senate in 1907. He proposed to establish four grounds to claim sovereignty in the Arctic region by Canada. He suggested the partition of northern lands based on geographical possession of a sector, so every country bordering on the Arctic region would extend its possessions up to the North Pole. Five sectors emerged: Norwegian (with Sweden), Russian, American, Canadian, and Danish. Opponents of the theory argued that it places the territory in the legal possession of a State even before it is discovered (Head, 1963, pp. 202–205). The United States rejected it while Norway and Denmark voiced some reservations. Canada used the sector theory as well with the doctrine of effective occupation in order to define part of the border. Based on it, in 1946, the Canadian government stated that a large part of the world's total Arctic is rightfully Canadian. It includes not only Canada's northern mainland but the islands and the frozen sea north of the mainland between the meridians of its east and west boundaries extended to the North Pole (Pearson, 1946). Russia accepted it and later the Soviet Union

9 The division of lands based on a meridian of longitude was used by States in the past.

incorporated it as a part of national policy to formally declare its claim to the Arctic in 1926. The current interpretation of international law indicates that the Sector Principle is more of a political proposition that a legal right (Shaw, 2003, pp. 455–456).

The US title to Alaska was never challenged. Canada's Arctic sovereignty was perceived vulnerable but in the end, Canada protected its title to the Arctic Islands in the 1930s by establishing Royal Canadian Mounted Police (RCMP) posts and surveying the area. Denmark's authority over East Greenland was challenged by Norway but the 1933 judgment of the Permanent Court of International Justice awarded Denmark sovereign authority over the entire island based on effective occupation (PCIJ, 1933, p. 86).

Closely paralleling evolution of the law of nations were the laws of the sea. Here the basic concepts of Roman law also prevailed, with *mare liberum* referring to high seas open to all nations and *mare clausum* or "a closed sea" under the authority of adjacent nations. In ancient Norway, royal sovereignty also applied to coastal waters, with the "median line" principle applied to marine boundaries between two countries. As a result, sea routes between Greenland, Norway, and Russia were considered to be a *mare nostrum*, meaning "our seas," and under the protection of Norway and Russia, a concept similar to the Roman *mare clausum* applied to the Mediterranean. Claiming a monopoly over the North Atlantic by the mid-thirteenth century, Norwegian kings banned foreign ships from trading north of Bergen, Norway's main port (Theutenberg, 1984, pp. 481–483). At that time, trade in Arctic goods such as furs and ivory was flourishing as a result of Greenland being incorporated into the Kingdom of Norway's "tax lands" or Crown lands. As demand for these items increased, ships from other European countries appeared on the scene. England and the Netherlands signed with the Danish/Norwegian Crown special treaties but tensions remained. In 1599, when the Danish/Norwegian Crown decided to uphold its claims by seizing several English ships, Queen Elizabeth I protested vigorously and invoked the law of nations in support of *mare liberum* (Theutenberg, 1984).

The law of nations was rudimentary in ancient times, as there was no means to enforce the principle of *mare liberum* without resorting to war. By the sixteenth century, economic and political factors forced transitional changes in laws applying to both land and seas. After gaining international support for the principle of "freedom on the high seas," countries like the Netherlands, France, and Britain focused on the "rights of innocent passage" through straits once claimed as territorial seas by adjacent nations. Eventually, the laws of the sea collided with the law of nations as competition intensified to lay claim to lands in the New World.

Nevertheless, since then modern international law of the sea has been developed based fundamentally on Hugo Grotius' treaties ("father" of international law) that advocated the principle of the freedom of the seas. Interpretations of laws of the sea have also evolved since the mid-twentieth century, particularly with regard to territorial waters, open seas, international straits, and the continental shelf. Most changes were incorporated into the 1982 United Nations Convention on the Law of the Sea (UNCLOS) that came into force in 1994. It provides a framework to determinate boundaries and legal rights in the Arctic Ocean. First of all, most of the Arctic Ocean falls under the jurisdiction of the five coastal States. Secondly, the Article 234 grants coastal States authority to regulate ice covered areas. Additionally, the coastal States have a right to extended continental shelf. The right of passage in international straits is also regulated but there is an issue with interpretation.

Exercising its inherent right as a major world power, the United States refused to ratify the convention without amendments that were still under consideration by the Senate under the George W. Bush administration. The United States has still not ratified the Law of the Sea, but has abided by its provisions in the Arctic region in the absence of ratification.

As we face new challenges resulting from global warming, historical definitions may become less relevant. Of greater importance are the patterns of cause and effect that attended changes in Arctic sovereignty and the characteristic response by each Arctic country to potential threat. Legal interpretations as they apply to Arctic sovereignty are still evolving and are still based on accepted principles set down in international law.

CONSEQUENCES OF CLIMATE CHANGE

Global Trends

> *All life on Earth is subject to the power of climate. Civilizations evolve or vanish forever. Favorable climatic conditions support the rise of great empires and promote trade prosperity and artistic achievement. Adverse climatic events often lead to war and other human catastrophes.*
>
> *(How Climate, 2015)*

There is an ongoing debate to what degree climatic changes are human-induced or whether they are simply following the natural course. Another raging issue is whether humanity should actually try to mitigate the changes or should they simply adapt to them. Should we take care of that or will Mother Nature do so, as she has always done? The facts, however, are merciless.

Our planet has heated up and is struggling to cope with global warming. For some time, scientists have been asking us to acknowledge these signs and change our course. We are at that point where we can decide what sort of climate we are aiming to achieve and want to have in the future. If we could mitigate climate change or reverse it, it would mean that for the first time the global society could control climate and ensure its stability for future generations, instead of climate controlling us and our destiny. All climate change affects life on Earth. Climate makes history – it always has and it always will.

Earth's climate system is highly interconnected which means that changes to the global climate influence all three countries both climatically and economically. Therefore, changes to ice sheet mass and energy flows in the far reaches of the planet affect our climate. "Life on Earth is sensitive to climate conditions; human society is especially susceptible due to the climate-vulnerable, complex, and often fragile systems that provide food, water, energy, and security" (The Causes, 2018).[10] Hence no State is an island, and therefore Canada, Greenland, and the United States are as susceptible to climate changes as the rest of the world, so all of them have their strategic, economic, and scientific interests vested in the Arctic.

"Everybody talks about the weather, but nobody does anything about it" is a famous American quote. The statement was allegedly made by Charles Dudley Warner but is usually attributed to Mark Twain (Quote Investigator, 2010). The question, however, is it really true that everybody talks about it and whether in reality nobody does anything about it. If we perceive it in a larger context and talk about the climate, and then take it for a metaphor, say, for the global warming, what impact could it have on the life on this planet and what action is expected of all of us to counteract its diverse and much threatening if not outright disastrous impact?

The climate has always been subject to large-scale but predictable fluctuations. Most of these are caused by the Sun, the center of our solar system. As the Earth revolves, the sun's rays strike its surface. At some times its orbit is

10 The data come from a seminar: The Causes and Consequences of a Rapidly Changing Arctic that took place on August 16, 2018 (Washington, DC, USA, and via webinar). The event was part of the Climate Science Special Report (CSSR) Seminar Series by the US Global Change Research Program in partnership with the National Oceanic and Atmospheric Administration. The presenter was Dr. Patrick C. Taylor, Climate Research Scientist, NASA Langley Research Center.

more circular, at others more elliptical. One orbital cycle takes 100,000 years. The angle of the Earth's axis also moves in a 40,000-year cycle. These changes caused regular climatic variations on earth. As the intensity of solar radiation increases and decreases, other influences on the Earth's climate such as fluctuations in sea/ocean currents become more irregular.

If we take the reader, however briefly, through a journey from 65 million years ago to the time of the Industrial Revolution, we may see the obvious: climate changes have had tremendous impact on the history of this planet and consequently on all its inhabitants, including humans. Climate has always influenced all life on Earth and the changes brought by its more serious alterations have shaped our planet. What we see and have today, perhaps not fully realizing it, is the outcome of the changes in the past. For example, it might be somewhat difficult to comprehend why the frozen land of today's Greenland bears this name as so little is green there, or why people once thrived in the area occupied today by the Sahara Desert.

Taken as a whole, the range of published evidence by the Intergovernmental Panel on Climate Change (IPCC) indicates that the net damage costs of climate change are likely to be significant and to increase over time (Parry, Canziani, Palutikof, Hanson, & van der Linden, 2007).

According to NASA studies,

> *...global climate change has already had observable effects on the environment. Glaciers have shrunk, ice on rivers and lakes is breaking up earlier, plant and animal ranges have shifted and trees are flowering sooner. Effects that scientists had predicted in the past would result from global climate change are now occurring: loss of sea ice, accelerated sea level rise and longer, more intense heat waves. Global climate is projected to continue to change over this century and beyond. The magnitude of climate change beyond the next few decades depends primarily on the amount of heat-trapping gases emitted globally, and how sensitive the Earth's climate is to those emissions. Temperatures will continue to rise because human-induced warming is superimposed on a naturally varying climate, the temperature rise has not been, and will not be, uniform or smooth across the country or over time. Frost-free season (and growing season) will lengthen, affecting ecosystems and agriculture. In a future in which heat-trapping gas emissions continue to grow, increases of a month or more in the lengths of the frost-free and growing seasons are projected. The increases will be considerably smaller if heat-trapping gas emissions are reduced.*

The IPCC provides a summary of key regional risks from climate change and the potential for reducing risks through adaptation and mitigation. Each key risk is characterized as very low to very high for three timeframes: the present, near term (here, assessed over 2030–2040), and longer term (here, assessed over 2080–2100). In the near term, projected levels of global mean temperature increase do not diverge substantially for different emission scenarios. For the longer term, risk levels are presented for two scenarios of global mean temperature increase (2°C and 4°C above preindustrial levels). These scenarios illustrate the potential for mitigation and adaptation to reduce the risks related to climate change. The projected impacts for world regions remind one a little of the ten biblical plaques, but they unfortunately are coming and unavoidable.

For the purpose of this book, let us examine the possible, or more accurately – predicted changes for Europe and North America (IPCC, 2014b, pp. 4–7):

(1) Europe. The climatic drivers include extreme precipitation and rising sea levels, warming, and drying trends with some extreme temperatures. The key risks list:

- Increased economic losses and people affected by flooding in river basins and coasts, driven by increasing urbanization, increasing sea levels, coastal erosion, and peak river discharges;
- Increased water restrictions. Significant reduction in water availability from river abstraction and from groundwater resources, combined with increased water demand (e.g., for irrigation, energy and industry, domestic use) and with reduced water drainage and runoff as a result of increased evaporative demand, particularly in southern Europe;
- Increased economic losses and people affected by extreme heat events: impacts on health and well-being, labor productivity, crop production, air quality, and increasing risk of wildfires in southern Europe and in Russian boreal region.

(2) North America. The climatic drivers include warming, high temperatures, extreme precipitation, damaging cyclones, and rising sea levels. The key risks list:

- Wildfire-induced loss of ecosystem integrity, property loss, human morbidity, and mortality as a result of increased drying trend and temperature trend;

- Heat-related human mortality;
- Urban floods in riverine and coastal areas, inducing property and infrastructure damage; supply chain, ecosystem, and social system disruption; public health impacts; and water quality impairment, due to sea level rise, extreme precipitation, and cyclone.

Outcomes for the Arctic

Much has been written about the changing climate in the Arctic. Actually, the very essence of discussing the Arctic is centered on climatic changes which will most probably lead to opening of new sea routes, allow access to new resources, impact the natural environment or at least significantly alter it, and change the lives of Indigenous Peoples forever, if they have not done it already. The Arctic is not only heating up but it is really hot at the moment as an international issue.

Much has been written about the importance of the Arctic Region, be it in the context of the climate and its global impact or the riches the Arctic might hold. After only a short glimpse at the map, the strategic importance of the region becomes obvious in terms of security. "The developments in the Arctic will have lasting ramifications for international relations, international law, international climate change policy, international trade, and international human rights. The Arctic's strategic location will not only influence government policies, but also affect strategic business decisions and global economic development over the coming decades. The Arctic Institute believes that an interdisciplinary approach to Arctic studies, bringing together researches and experts from a variety of disciplines, is key to fully understanding the ongoing transformation of the region and making progress towards sustainable Arctic development" (Arctic Institute, 2013).[11] The truth is that the whole Mendeleev periodic table can be found in the Arctic, not to mention abundant fisheries, new maritime passages, and incredibly rich wildlife. After all, considering maritime life alone, there must be a reason why whales migrate up north to find the richest acqua hunting grounds. In short, there is no shortage of reasons why the Arctic is drawing unprecedented international interest and

11 "The Arctic Institute seeks to establish itself as an authoritative, interdisciplinary, and independent source for information and in-depth analysis about the developments in the High North. The Institute was founded in 2011 and currently aims to bring together scholars and researchers to build a growing stock of knowledge and expertise on the Arctic region. In contrast to existing platforms for Arctic affairs, The Arctic Institute is not affiliated with or sponsored by any of the Arctic states" https://www.thearcticinstitute.org/.

becoming increasingly important to the world's most powerful countries and their economies. But the climatic changes so far seem to reign supreme.

> *The Arctic now sees shorter periods of crushingly cold weather and, in some areas, longer stretches of warmer weather. In mid-February 2018, temperatures at the world's northernmost weather station were above freezing – some 45 degrees Fahrenheit above normal. Meanwhile, the thickness of Arctic Ocean sea ice declined by more than 65 percent over the past 30 years, according to a 2017 Arctic Council report. And open water will absorb the sun's glare, rather than reflect it. This will probably lead to warmer temperatures and further melting.*
>
> *(McFarland & Lide, 2018)*

The experts from IPCC have the following to say about the possible socioeconomic impacts in Polar Regions (Arctic and Antarctica):

> *Climate change will have costs and benefits for polar regions. Climate change, exacerbated by other large-scale changes, can have potentially large effects on Arctic communities, where relatively simple economies leave a narrower range of adaptive choices.*
>
> *In the Arctic, positive impacts include new possibilities for economic diversification, marine shipping, agricultural production, forestry, and tourism. The Northern Sea Route is predicted to have up to 125 days per year suitable for navigation by 2050, while the heating energy demand in the populated Arctic areas is predicted to decline by 15 percent. In addition, there could be greater accessibility to offshore mineral and energy resources although challenges related to environmental impacts and traditional livelihoods are possible.*
>
> *Changing sea ice condition and permafrost thawing may cause damage to bridges, pipelines, drilling platforms, hydropower, and other infrastructure. This poses major economic costs and human risks, although these impacts are closely linked to the design of the structure. Furthermore, warmer winter temperatures will shorten the accessibility of ice roads that are critical for communications between settlements and economic development and have implications for increased costs. Statistically, a long-term mean increase of 2°C to 3°C in autumn and spring air temperature produces an approximately 10- to 15-day delay in freeze-up and advance in break-up, respectively.*

> *Changing sea ice conditions will impact indigenous livelihoods, and changes in resources, including marine mammals, could represent a significant economic loss for many local communities. Food security and health and well-being are expected to be impacted negatively.*
> *(IPCC Assessment Report, 2014, p. 1595)*

The evidence seems to be irrefutable, but there are still some people who refuse to believe or at least acknowledge it. However, it is not a matter of beliefs, but pure science. People may still go on believing that the Earth is flat and the Sun revolves around it, and that evolution is just a figment of somebody's imagination. If they are set in their beliefs, there is nothing to be done as science can hardly compete with faith or beliefs. It seems, nevertheless, that it is better to know because such an awareness could lead to two basic ways of approaching the problem: mitigation and adaptation.

As highlighted by the IPCC, there are several key climate-related risks in the Arctic and Antarctica, and potential adaptation practices.

The key risks include:

- risks for freshwater and terrestrial ecosystems (high confidence) and marine ecosystems;
- risks for the health and well-being of Arctic residents;
- risks for northern communities due to complex interlinkages between climate-related hazards and societal factors, particularly if rate of change is faster than social systems can adapt.

Potential adaptation procedures include:

- improved understanding through scientific and indigenous knowledge;
- enhanced monitoring, regulation, and warning systems;
- hunting or fishing for different species, if possible, and diversifying income sources;
- coproduction of more robust solutions that combine science and technology with indigenous knowledge;
- improved communications, education, and training;
- shifting resource bases, land use, and/or settlement areas;
- adaptive comanagement responses developed through the settlement of land claims (IPCC, 2014, p. 1594).

POLITICAL AND ECONOMIC PERCEPTION OF THE NORTH AMERICAN ARCTIC

Political Setting of the North American Arctic

In the nineteenth century, there was a major shift in sovereign authority in the North American Arctic. Denmark separated from Norway, but retained control of its Greenland colony, and the United States purchased Alaska from Russia, followed by Canada's purchase of land rights held by the Hudson's Bay Company and finally there came Britain's transfer to Canada of its discovery claims to the Arctic Islands. Since the twentieth century, the North American Arctic has been very much present in international politics.

Although initially perceived as an area of the Cold War competition, after the changes in the geopolitical situation, it became a subject of scientific cooperation and concerted efforts directed at environment protection. Currently, the international importance of the Region seems to be undergoing a significant transformation. Politically, the region is divided between Canada, Denmark/Greenland, and the United States.

> *The consequences of the climatic changes occurring around the North Pole and the accompanying it increased activity of the States possessing territories beyond the Northern polar circle have contributed to the significant growth of interest in this area by the international community. The attention has been focused both on the problem of possible negative consequences of the new situation, and the visions of substantial benefits which may be drawn from the so-far inaccessible Arctic.*
>
> *(Reflections, 2010, p. 692)*

Such statements are hardly surprising if one realizes that the expected access to a wealth of raw materials and resources and the monitoring of Arctic sea routes have resulted in both a reason for competitive efforts, and an impulse for the search of new forms and mechanisms of international cooperation in the High North. There exist many compelling reasons justifying the problems of international relations in this region to become a subject matter of joint analyses and scientific cooperation. This, in turn, requires close cooperation of scientists, in conducting active research in the Arctic both in the fields of natural sciences, as well as in social sciences.

The Canadian North gained its strategic value with arrival of European fur traders. The Hudson's Bay Company, established in 1670, controlled the fur trade for several centuries. In 1869, the Hudson's Bay Company transferred all its land to the British Crown and then the lands were turned over to a newly formed Dominium of Canada. Since then, Canada has steadily affirmed its occupation of the North. The North came under the control of federal government through territorial governments. Today, 40 percent of Canada's landmass is located north of the 60th latitude and 162,000 km of Canadian coastline is in the Arctic. This region is divided between three territories: Yukon, Northwestern Territories (NWT), and Nunavut, and central to Canada's national identity and values. It is visible not only in politics but also deeply rooted in the culture. Aritha van Herk (2018, p. 53) argues that

> *To Canadians, 'the north' is larger than a geographic space, but performs as an imaginary and imagined realm of idea and inspiration. The cultural influence of this all-encompassing idea and area goes far beyond its dimensions to measure Nordicity, intimacy, and directionality.*

Daniel Chartier (2018) points out two sociopolitical phenomena that had an effect on the representation and the reception of the North and the Arctic. On the one hand, the general context of indigenous colonialism, which reinforced the silencing of cultural and human aspects of cold territories, and on the other hand, the general tendency of the governance of the "North," is dominated by the capitals or the powers of the South, who administrate according to their knowledge (seldom based on experience) and the circumstances of their own needs, with the gaps that they can create. The Arctic and especially "Arctic sovereignty" became a key phrase in political discourse during Canadian Prime Minister Stephen Harper's (2005–2015) decade. The North was viewed as a symbol of a positive nationalism and proud heritage that needed to be protected and developed. The declaration that the Arctic is fundamental to Canada is still in place.

Denmark is considered an Arctic State because of Greenland. Greenland covers an area of 2,175,600 km^2 of which 1,833,900 km^2 are lands covered by a permanent ice sheet. Only 150,000 km^2 can be considered inhabited land. Greenland is considered an island, although discussions about its continental character are not unjustified. While determining whether a territory is an island, two factors appear to be decisive.

Firstly, it needs to be isolated and Greenland certainly is. Secondly, it should be relatively small and Greenland certainly is not. However, Matt Rosenberg explains that the decisive factors are: the fact that Greenland belongs to the American tectonic plate, it lacks unique culture or the flora and fauna, and the Greenlanders consent to call themselves islanders. Geographical arrangement indicates that there are dependencies between Greenland and the

American continent, that is the United States and Canada, while there are no such dependencies with the European continent and thus with the Kingdom of Denmark. As already mentioned, Greenland by its location and geological history is closely connected with the American continent. Climatically, on the other hand, it belongs to the Arctic, because as much as 85% of its surface is covered with ice. The location of Greenland and its size extend these Arctic areas to the south, which is the biggest climatic anomaly in the northern hemisphere (Tomala, 2016, pp. 48–51).

So it is Greenland that makes Denmark an Arctic State, though the latter is located in the temperate climate. Greenland is the coldest territory of the Kingdom of Denmark. Although the relations between the two subjects of international relations have been loosening in recent years, they are still linked by a common foreign policy, including issues related to governance in the Arctic (see Chapter 2).

The United States, as is most probably commonly known, became an Arctic country through the purchase of Alaska from Russia on March 30, 1867, for a price of $7.2 million; the "best deal ever" was executed by US Secretary of State William H. Seward. Not too many Americans realize that, by virtue of Alaska, the United States is indeed an Artic country. What is even more surprising, not too many seem to care. There are many good reasons why the United States is called a reluctant Arctic country.

The satellite picture in Fig. 1.1A, on the right, presents the Bering Strait (international waters) separating Russia and the United States: the arrow in Fig. 1.1B points to the Russian Big Diomede; very close to it, the American

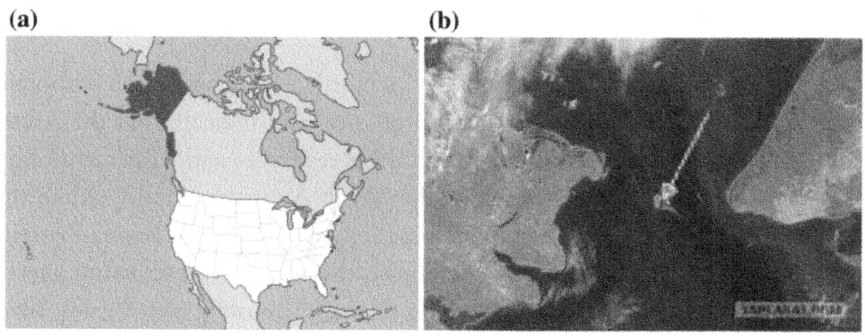

Source: (A) https://www.nationsonline.org/oneworld/map/USA/alaska_map.htm. (B) http://www.yaplakal.com/.[12]

Fig. 1.1. (A) Picture on the Left-hand side: The United States and Alaska (in Red). (B) International Waters Separating Russia and the United States.

12 Interestingly, the Russian website "yaplakal" means in English "I cried" or "I wept" which might show some Russian sentiment toward the Alaska sale.

Little Diomede is situated. At their closest points, the two islands are about 3.8 km (2.4 mi) apart. If one could handle the cold water, one could possibly swim from the United States to Russia, or, depending on the fancy, perhaps the other way round!

Naturally, as everything else, such a deal must have had its enthusiasts and staunch opponents. The opponents from years ago were the ones that coined such names for Alaska as "Seward's folly," "Seward's icebox," and President Andrew Johnson's "polar bear garden," while some senators called it the procurement "reckless and wasteful." The proponents call the purchase the deal of all times and praise Alaska's beauty and natural resources to no end. Clearly, to nobody's surprise, the matter has two sides.

> *The territories located in the North American Arctic, when viewed through the lens of their "mother" countries, seem to be light years apart. However, each circumpolar culture is the product of a synthesis of two or more cultures, from the South and from the North. Multicultural interactions thus define the North and the Arctic. Depending on the place, there is more or less diversity, more or less harmony between those of Indigenous and non-indigenous origins. The Greenlandic identity, for example, is today a synthesis of several centuries-old Inuit cultures combined with those of the missionaries, the Danish colonizers, and recent immigration.*
>
> *(Chartier, 2018, pp. 47–48)*

Highlights of the North American Arctic Economy

There is a lot of economic potential in the North American Arctic. Although oil and gas reserves are still unknown, the estimates indicate that the Arctic comprises 30 percent of the world' natural gas resources and 13 percent of oil reserves (USGS, 2007). Nevertheless, the North American Arctic, unlike the well-developed Nordic Arctic and the increasingly redeveloped Russian Arctic, is far behind its Arctic neighbors. Foremost, there is an infrastructure gap in the North American Arctic which relates to roads, housing, ports, grids, fiber optic cable, and other necessary components (Shadian, 2018). As it is estimated by the World Economic Forum, the Arctic needs at least $1 trillion for the investment in infrastructure (Roston, 2018).

In the Canadian part of the North American Arctic, oil and gas deposits are found around Beaufort Sea at Makenzie Delta and in the Canadian Arctic Archipelago. So far, since 1999 gas has been extracted from Ikhil gas field at the Beaufort See for the town of Inuvik (NWT) and oil in Norman Wells (NWT). Additionally, the deposits of other minerals resources including gold-silver, nickel-copper, diamonds, and coal have been discovered in the Canadian Arctic (Government of Canada, 2020). For example, in 2018, Canadian rough diamonds production (Canada was the world's third largest diamond producer by value and volume) came from five mines located in the North. Additionally, 6.1 percent of Canadian copper manufacture, 7.6 percent of gold, and 9 percent of iron ore were produced in the North (Government of Canada, 2020).

The potential economic benefits of extraction industry in the Canadian Arctic are constrained by infrastructure. Everybody is aware of housing shortages, high food prices, and insufficient access to health care, but what really halts the development is a deficiency of the means of transport.[13] The land and water modes of transport are limited so air travel prevails but with gravel runways. So even if the extraction industry develops, the shortage of transport infrastructure might reduce its competitiveness. The next important factor influencing the economy of the region is the shortage of qualified labor mainly due to limited access to higher education and the Internet. The Business Index North 2018 report noted that many cities in the Arctic were confronting challenges stemming from the loss of the region's youth – who move south in search of education and jobs – and a gender imbalance (Gosnell, 2018).

There is manufacturing and quarrying, and oil and gas extraction, but this is not the primary source of revenues of three Canadian territories (Statista, 2020). So far, the territories have problems in generating revenues. The territorial governments are heavily dependent upon subsidies from the federal government. In 2018–2019, the Government of Nunavut received $1.6 billion in major transfers, accounting for 93 percent of its revenues in this fiscal year (Table 1.1).

Table 1.1. Federal Support to the Territories in 2018/2019.

Territory	Total Federal Support (Million Dollars)	Support per Capita (Dollars)	Federal Transfers % of Revenue
Yukon	1,008	24,898	81
NWT	1,321	29,227	80
Nunavut	1,633	42,985	93

Source: Federal Support to Provinces and territories, Department of Finance Canada for 2018/2019.

13 The social conditions of the Canadian North are well researched. They can be found in several reports and strategies, especially in the 1996 Report of the Royal Commission on Aboriginal Peoples. The 2019 Canada's Arctic and Northern Policy Framework also identifies them and looks for remedies. See Chapter 2.

Economic outlook does not look good for NWT as the exploration investments in the NWT have stalled for the last 12 years (Territorial Outlook, 2019) what might influence on declining population over the last five years (Statistics Canada, 2019). There is mineral wealth in Nunavut, but high cost of production and transport and shortage workforce in the resident's population slows back the extractions (Territorial Outlook, 2019). As the Executive director of the NWT and Nunavut Chamber of Mines Tom Hoefer indicated, the mining companies have to face many challenges in the region. Operating in Canada often means creating own transportations system, building roads, seaports, and airports and providing own power either diesel generated or wind (Klein, 2018). This increases the cost of production.

Overall, Canada's Arctic economic development is falling far behind in comparison with the other northern regions in other parts of the world. John Higginbotham of the Centre for International Governance Innovation suggested that in order to improve the economic development the Canadian North should start forming alliances with groups in Alaska and Greenland. By forming a united front, they will be stronger in attracting the private sector (Weber, 2018).

Greenland economically is strongly dependent on Denmark, which contributed to its underdevelopment for years. Difficulties occurring in Greenland are connected with functioning in isolation from Copenhagen, with an islandic location and very small number of inhabitants. The indicated factors cause financial and economic reliance typical of a dependent territory. Greenland after World War II, despite separatist moods among the citizens, by the decision of the local parliament committee decided to maintain dependence on the Danish Kingdom. It did not make efforts to regain independence. Such a decision of the community could be questionable if it was not for the fact that Greenlanders were dependent on financial aid from Denmark and were not able to manage their own territory themselves.

At that time, the official goal of the governmental policy of the Kingdom was to provide the people of Greenland with similar development opportunities as it was in the Realm of Denmark. There are doubts about the way of redistribution of this aid, which in the opinion of Greenlandic politicians, made Greenland dependent on the Kingdom of Denmark. R. Gessain (1978) points out the examples of anomalies in this respect, occurring after World War II, which made Greenland's economy dependent on the Kingdom – thus not allowing it to become more competitive in the international arena. Among many examples, the issue of employment in Greenland draws attention. It did favor workers from Denmark. Theoretically, the Danes were under the illusion that all Greenlanders should be given a job. Danes still in 1966 delivered all prefabricated houses from Denmark and also sent workers to the island to assemble them. This was explained by the lack of professional preparation of the Inuit. The answer to the question: "Why wasn't it given to them?" was: "According to trade union standards, they would have to undergo four years' technical training in Denmark." Another example of the controversial Danish activity was arbitrarily imposed housing policy that forced the Greenlanders to move out from their scattered settlements to concentrated ones, thus drastically changing their lifestyle (Gessain, 1978).

In assessing the effectiveness of Danish aid, attention should be paid to the social and economic consequences of the policy. Greenlanders consume about 14.7 billion DKK per year, while their own domestic production equals only 5 billion DKK (Fig. 1.2).

As the figure shows, consumption in this area is three times higher than production. Such an imbalance has an unfavorable influence on the Greenlandic economy, strongly dependent on the Metropolis. Without the help of the Danish Crown, currently Greenland would not be able to reconcile the local budget. So it is not surprising that for years most of the aid, which was provided to Greenland, did not really leave Denmark, but at the same time gave it control in many sensitive areas, such as over natural resources. Paradoxically, the hope of Greenlanders for independence is global warming and increasing GDP without Danish support.

Since the end of World War II, Greenland has been quite cautious about independence due to its weak economic status (Fig. 1.3).

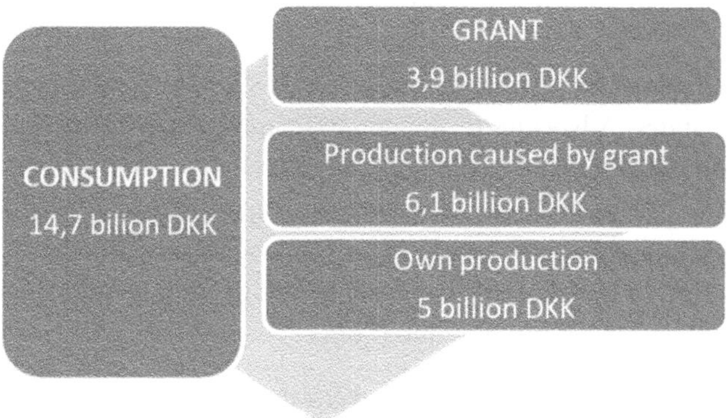

Source: Elaborated by M. Tomala on the basis of: M. Holmgren, E. Ronnle, Hydrocarbons in Greenland. Prospects for the Greenlandic Economy, Lund 2012, p. 18.

Fig. 1.2. Greenlandic Consumption (Disposable GNI), Production (GDP), and the Block Grant including Danish State Expenditure in Billion.

Source: Elaborated by M. Tomala on the basis of D. Michelsen (ed.), Greenland in Figures, 2018 In: http://bank.stat.gl.

Fig. 1.3. Gross Domestic Product in Greenland in Years (in DKK Billion).

According to the World Bank, the GDP of Greenland is constantly growing. Due to a small number of people, GDP per capita seems to be a better indicator positioning Greenland, and it amounted to 48,181 USD in 2016, which classifies Greenland quite high. The GDP per capita is similar to the average for European economies. It should be remembered that the economy is critically dependent on significant support from the Danish government, which provides about half of the income. Greenland has a one-sided production and demands a wide range of goods, which necessitates extensive foreign trading. Apart from fishing and hunting, Greenland has a very limited domestic production of commodities. This causes import to include more or less all goods necessary in households, businesses, and institutions. As the majority of Greenland's exports consist of fish and shellfish, Greenland is very dependent on the fishing sector and the international buyer prices.

One could analyze the Alaska purchase from many varying points of view. The financial aspect of it, naturally, comes to the front and should be examined first.

The land added 586,412 square miles (1,518,800 km^2) of new territory to the United States. Advocates of the purchase immediately claim that the price was a mere $7.2 million ($123 million today), or about 2 cents per acre ($4.74/km^2). The quoted figures are constantly questioned and challenged, but 2 cents an acre does seem to be a bargain. Some scholars claim that Alaska was clearly overpriced and that the real value of it at the time should have stayed at about 5 million, which was the initial offer.

> *Senator Gwin tendered a hypothetical offer of five million dollars for the Russian colony, a figure Gorchakov (the then Minister of Foreign Affairs) found far too low. (Eduard) De Stoeckl (the Russian minister to the United States) informed Appleton (Undersecretary of State John Appleton) and Gwin of this, the*

> *latter saying that his Congressional colleagues in Oregon and California would support a larger figure.*
>
> *(Bolkhovitinov, 1990)*

Several sources suggest that some of the money never made it to the Russian coffers and was used to bribe American Senators to approve the price (Barker, 2009, p. 36).

Those in favor of the purchase say that the deal has paid for itself many times over.

> *Hundreds of billions of dollars in whale oil, fur, copper, gold, timber, fish, platinum, zinc, lead and petroleum have been produced in Alaska over the years – allowing the state to do without a sales or income tax and give every resident an annual stipend. Alaska still likely has billions of barrels of oil reserves.*
>
> *(Hensley, 2017)*

The gold discovered in the Yukon Valley alone during the Gold Rush was worth billions and it must be noted that the state mines gold to this day.

Naturally, there are many schools of thought on the matter and several methods of calculating the gains and losses. Economist David Barker from the University of Iowa claims that after adjusting for the relative size of the national economy, then and now, the price tag in today's dollars would be a staggering $16.5 billion. As always, for every argument there is a counterargument. Many scholars simply question the method of calculation used by Barker.

The very beginnings of American interests in Alaska were not connected with exploiting the natural resources of the region but definitively with a greater scheme, i.e., with expanding the American economy. According to Stephen Haycox, William Henry Seward, an expansionist just like many other American expansionists before him, to name only Thomas Jefferson and Kohn Quincy Adams, vigorously sought new markets for American products. Seward was of the opinion that the future for American business lay in Asia (Haycox, 1990, p. 65).

Alaska is extremely difficult and very expensive to govern for a variety of reasons. Distances are vast and conditions are difficult. Building reasonable infrastructure has been very tricky and extraordinarily expensive. Ordinary government expenditures, such as the cost of running a territorial legislature, have been much higher in Alaska than elsewhere. Ambitious projects tend to run considerably over budget (Barker, 2009, p. 20). Various experts maintain that Alaska has been receiving more from the federal government than it has ever contributed to the budget as the revenues generated in Alaska stay mostly there, i.e., in the state. One cannot help making a comparison of the situation with Denmark and Greenland – what is it going to be: subsidies forever or the future exploitation of the riches? (Czarny, 2015, pp. 120–124). The aforementioned Barker (2009, p. 3), a staunch critic of the financial viability of the Alaska purchase, claims that

> *If the federal government made a substantial profit from Alaska, as is generally believed, then Alaska would provide a counter-example to this claim. A financial loss for the federal government would support the proposition that Alaska has been dependent on the federal government.*

Not that he is alone as there are many critics of the "wasteful spending," the more so as many Alaskan residents

> *...seem to revile the federal government even as their politicians excel at reeling in and spending its money. Alaska is the top recipient of federal stimulus dollars per capita – with no close second. Alaska, as a new state with vast needs, required ports and airports and highways and so on. But then you must account for the fact that Alaska's share of federal spending has spiked sharply in the last 15 years.*
>
> *(Powell, 2010)*

The Polish sources seem all to praise the deal, for whatever reason, with a notable exception of a few voices. The titles of the Internet posts 150 years after the purchase, in translation, leave no doubt as to the assessment of the deal: The worst deal ever or how the Russians sold Alaska to the Americans for a song (Strawiński, 2017), The worst deal in Russian history – the sale of Alaska in 1867 (Balcerowski, 2015), The worst deal in history: How the Russians sold Alaska (Osiński, 2016).

There are also some dissenting voices, mostly based on the attitude and reasoning of the aforementioned economist David Barker: The Alaska purchase – the worst deal in America's history (Woś, 2012), The purchase of Alaska by the United States: The worst investment in the American history? (Smolińska, 2016).

> *A strictly cash flow-focused analysis of Alaska's impact on the federal treasury since the Treaty seems to indicate that all else excluded, the purchase was not a good deal. While there are, of course, countless counterarguments to show that states, and most*

> *large-scale public infrastructure investments, rarely stand up to such scrutiny, perhaps a most important point missed by such calculus is the savings on what wasn't. The geopolitical externalities from a Russian Alaska, had it remained in place into the 21st century, would likely be tipping the scales in favor of the purchase.*
>
> *(Blair, 2017)*

CONCLUSIONS

This chapter delineated the stage for the application of the Nash game theory, which is the North American Arctic. The climate change made the Arctic visible, hence the North American Arctic as well. With the quest for natural resources and shipping routes intensified, the North American Arctic States has focused more on the region and the people living there.

The Arctic appears to have all the makings of approaching profound changes with consequences of unimaginable magnitude for the planet Earth. Scientists have tried to project, describe, and prove the possible impacts in order to present to people what awaits us in the not too distant future. Not that it has worked very well. For every proponent, there is an opponent. Some quite accomplished movers and shakers of this world with a lot of power and influence, notably including world politicians, outright question the very idea of global warming, claiming it is a scheme against humanity. In spite of available evidence, the body of knowledge already gathered, they refuse to acknowledge the fact that sometime soon we may face a crisis similar to those described in "How Climate Made History." All the signs are right in front of our eyes.

And it is not only about the Arctic, but the whole planet. The proverbial saying: "What happens in the Arctic, does not stay in the Arctic" seems to hold true. As weather systems are chaotic in the sense discovered by Edward Lorenz (1917–2008), "The flap of a butterfly's wings in Brazil can set off a tornado in Texas (Lorenz, 1972)." Therefore, we are all going to be affected, and to a degree, we have already been affected by what is currently happening in the Arctic.

The authors wish to make an important disclaimer here: although our planet is vulnerable to climate changes, which at times had proved instrumental, we do not necessarily believe that the series of historical events in the history of mankind are a simple cause-event chain and should be all attributed to those events. There had always been also many other factors at play.

2

LEADING PLAYERS IN THE NORTH AMERICAN ARCTIC

INTRODUCTION

The governance in the North American Arctic is unique. The power is spread vertically and horizontally between many levels of government. The constant interaction occurs between local, subnational, and national levels. It is worth to note that since the end of twentieth century, local authorities have gained more control over the state of affairs in the region. Especially governance on the subnational level in the North American Arctic has been exceptional in the last 50 years. Due to the process of devolution, subnational levels have received more power. Additionally, self-governments of Indigenous peoples have been gradually included in the decision-making process that requires a horizontal cooperation between various local institutions. In the Canadian part of the North American Arctic, the three territorial governments, with self-governments of Indigenous peoples, shape the subnational governance. Greenland is a prime example of almost complete devolution of power. Nevertheless, politically it is still a part of Denmark, as a subnational unit that also represents that indigenous population. Governance on the subnational level in the American part of the Arctic is primarily in the hands of the State of Alaska, although the Indigenous peoples also exercise certain authority. On the national level, the North American Arctic States developed various strategies and policies toward the Arctic based on national objectives and international developments. Within the last 10 years, Canada launched two intertwined strategies in 2009 and 2019, and Denmark presented its 2011–2020 Strategy in 2011. Also the administration of Barak Obama released an official Arctic policy only in 2013 which was something quite new in American politics, as previously all that was available were Presidential

Directives. Current policies toward the Arctic are primarily based on objectives and principles emphasized in the strategies.

SUBNATIONAL AND SELF-GOVERNMENTS

Historical Background

The territories were admitted to Canada at the end of the nineteenth century. The federal government bought the Rupert's Land and the North Western Territory in 1870 and acquired the Arctic islands and part of the Yukon 10 years later. By 1880, the Northwest Territories (NWTs) comprised all lands and waters in present-day Canadian North.[1] Later, the NWT was divided. Yukon turned into a distinct territory in 1898 and then the southern part of territories below the 60th parallel of latitude separated. Thus, the southern boundaries of the territories were set at this parallel at the beginning of the twentieth century. The third territory, Nunavut, was carved out of the NWT along the ethnic line where the Inuit make up 85% of the population. Today, there are three territorial governments in Canada: Yukon, the NWT, and Nunavut. The territories include all of mainland Canada north of latitude 60 north and west of Hudson Bay, as well as all Arctic islands. The Canadian Constitution of 1867 did not assign any power to the territories which fell under the federal jurisdiction. Initially, they were administered by the federal government. A commissioner appointed by and responsible to the federal government oversaw the administration. Gradually, the power of commissioner had been reduced and governments responsible to elected legislative assembles instituted. The commissioners still serve as federal governments' representatives, but their role has become mostly ceremonial.

Greenland was discovered in the tenth century by the Norwegian, Erik the Red, who established a small settlement there. The settlement did not survive, but when the European whalers' expeditions reached Greenland, they met the Inuit. Then, in 1721, the missionaries of the joint Kingdom of Denmark and Norway settled there and converted the Inuit to the Christian faith. The joint kingdom established sovereignty over the territory. When Norway and Demark separated in 1814, the Treaty of Kiel secured the Kingdom of Denmark's possession of Greenland. This was later unsuccessfully contested by Norway before the Permanent Court of International Justice in The Hague in 1933.[2]

1 Except for a small part of Newfoundland and Labrador that joined Canada in 1949.

2 Legal Status of Eastern Greenland, PCIJ Series A/B. No 53.

Thus, Greenland became a Danish colony. During WWII, the Danish government signed an agreement with its US counterpart to hand over control over Greenland to the US in exchange for the right to construct air and naval base on the island. On April 10, 1941, the United States established a protectorate over Greenland. This war time presence of the US army in Greenland resulted in construction of several military installations, including the still operating Thule Air Base. After the war, the American administration made an offer to buy Greenland, but the Danish government refused to sell. From 1945 to 1953, Greenland was regarded a Danish colony. In 1953, under the Constitution of Denmark, Greenland became an integral part of the Kingdom of Denmark as a district, with two seats in the Danish Parliament.

As already indicated, Alaska became a part of the United States on March 30th, 1867, based on the agreement to purchase Alaska from Russia for a price of $7.2 million. It is estimated that by the time of the cession, only some 50,000 Indigenous people were left, together with 483 Russians and 1,421 Creoles (descendants of Russian men and indigenous women). Alaska became a state in 1959, when President Dwight D. Eisenhower signed the Alaska Statehood Act, allotting it 104 million acres of the territory (Alaska Statehood Act, 1958). Before, Alaska had a status of just a territory, with major legal consequences.[3] Eisenhower's Statehood Act

> *...contained a clause emphasizing that citizens of the new state were declining any right to land subject to Native title – which by itself was a very thorny topic because they claimed the entire territory.*
>
> *(Hensley, 2017)*

Powers and Responsibilities of the Subnational Governments

Powers and responsibilities of the subnational governments are the product of different political agreements with the state governments based on political, social, and economic circumstances, so they vary.

3 First, Alaska was a department with no civilian government, then a mining district, and after that, in 1912, it gained a territory status. The territorial system was designed by Congress for areas such as the Louisiana Purchase, the Oregon Country, or Alaska that wanted to be part of the United States, but these regions had to prove that the area had a large enough resident population and a way to sustain itself economically. Until then, Congress would hold those areas in a sort of second-class status, as a territory. In territories, the governor was not elected; he was appointed by the US President. At first, territories had no legislatures. Then, when Congress authorized it, the resident citizens could elect a two-house legislature, a Senate and a House but with limited power, controlled by Congress.

Each of the Canadian territories has a distinct set of differences. The NWT is the most populous, where the Indigenous peoples (Indian, Inuit, and Metis) constitute half of the population of 45,000.[4] The Government of the NWT works under the consensus system. The Executive Council is the major decision-making body responsible for the overall management of the NWT. Along with the federal and territorial governments present in the NWT, there are also seven indigenous regional governments and four indigenous community-based governments.[5,6] Yukon is the smallest and the least populated territory with the majority of population of European descent. The Indigenous peoples account for 25% of the total 41,000 inhabitants (Statistics Canada, 2020). Since 1978, Yukon has also adopted the principles of responsible government. The Premier is responsible to the elected Legislative Assembly in elections organized along political party lines. Thus, Yukon is the only territory in which political parties compete for power. Nunavut has the government that works also under the consensus system. Similarly, to the NWT, all members of the Legislative Assembly are elected as independents in their constituencies every four years. As already mentioned, the federal government delegated various powers to territorial governments. The jurisdiction of the NWT's government greatly expanded in 2014. The devolution agreement gave the NWT more control over land and resources (Northwest Territories Act, 2014). According to the agreement, the territorial government gained jurisdiction over 90% of land and water. The rest of the area that include national parks remained under the federal control. Environmental assessment and overlapping oil and gas remained shared responsibility. The NTW's government tries to attract more investments not only in oil, gas, and mining industry but also in other sectors such as tourism, fishing, or manufacturing (Northwest Territories Strategy, 2020). So far, the NTW economy has depended on diamond, metal mining such as copper, and the federal government funds. Economic outlook does not look good for NWT; the exploration investments in the NWT have stalled for the last 12 years (Territorial Outlook, 2019) which might have influenced the declining population over the last five years (Statistics Canada, 2019). Yukon devolution agreement that entered into force in 2003 provided for the transfer from Canada to the Yukon Territorial Government

4 Indigenous peoples of Canada include Indians or First Nations, Inuit, and Metis.

5 Akaitcho Territory Government; Dehcho First Nations; Gwich'in Tribal Council; Inuvialuit Regional Corporation; Northwest Territory Métis Nation; Sahtu Secretariat Incorporated; Tłįchǫ Government.

6 Kátł'odeeche First Nation; Salt River First Nation, Acho Dene Koe First Nation, and Fort Liard Metis Local #67; Délįnę Got'įnę Government.

of responsibilities for lands, water, forestry, and mineral resources (Yukon Act, 2002). This agreement, also signed by the Indigenous people, gave Yukon legislature the power to make laws with respect to public land and waters. In 2012, amendments changed revenue sharing arrangements providing the Yukon government with more revenues from the natural resources industry. The issue of adequate revenue is very important for the territory, so the government looks for a new potential. So far, residents are mainly employed in public administration. There is also manufacturing, mining, and oil and gas extraction, but it is public administration, real estate, and construction that mainly contribute to the GDP (Statista, 2020). Following the debate in Canada about Aboriginal self-government, the Inuit from Nunavut signed two interlinked agreements: the Nunavut Land Claims Agreement (NCLA) and the Nunavut Act that received royal assent on June 10, 1993. The second agreement fulfilled the commitment stated in Article 4 of the NCLA to establish a new Nunavut territory with its own Legislative Assembly and public government. Nunavut came into being on April 1, 1999 through the division of the NWT. Nunavut became the largest territory that encompasses the traditional land of Inuit, which constitute around four-fifths of 41,000 inhabitants. It has young and fast-growing population. The devolution agreement is still a work in progress in Nunavut. So far, the federal government has signed an agreement in principle with the government of Nunavut on devolving legislative powers in relation to land and resource management which is a base toward a final devolution agreement. There is mineral wealth in Nunavut, but high cost of production and transport and shortage of workforce in the resident's population slows back the extractions. However, a growth in gold mining has been observed (Territorial Outlook, 2019).

Hopefully, devolution and self-governments will bring more financial responsibility and social stability to the North American Arctic. So far, the Canadian territories have problems in generating revenues and are heavily dependent upon subsidies from the federal government, especially Nunavut. In 2018–2019, the Government of Nunavut received $1.6 billion in major transfers, accounting for 93% of its revenues in that fiscal year (Government of Nunavut, 2019). Barry Dewar indicates that the Government of Nunavut finds itself in the middle of a bilateral treaty relationship between the Inuit and Canada, where it has major implementation responsibilities, but where it is not recognized as a full partner and where it does not feel adequately resourced to fulfill its treaty obligations or the expectations of the Inuit (Dewar, 2009). Yukon and NTW governments also received the federal financing for 80% of the budgetary expense. Social stability is impacted by cultural transformation, abuse in residential schools, and discrimination.[7] As a result, Indigenous people have difficulties in accommodating the changes. Substance use and alcohol misuse have been identified as a primary health and social concern. Overall, the territories have highest rate of alcohol and drug-related hospitalization (Hwang, 2019). To tackle these obstacles, the territories commenced the regional cooperation by signing agreement on Pan Territorial Vision for Sustainable Development in 2017, which also includes collaboration with local indigenous governments and organizations. This agreement's primary objective is fostering long-term development of northern economies on a foundation of responsible resource development, economic diversification, and job creation (Government of Nunavut, 2017).

Following the changes in the Greenlandic society in the 1960s, the process of devolution of power to a Greenlandic self-government commenced. First, following negotiations by the Danish–Greenlandic Home Rule Commission, the Danish Parliament passed the Greenland Home Rule Act that proposed the changes in governance by introducing democratic institutions in Greenland. In a referendum held in 1979, the majority of the Greenlandic population (73%) accepted the principles of home rule which established a government responsible to Greenlandic legislature. The Act states that Greenland constitutes a distinct community, a society of people in the Kingdom of Denmark with the home rule authorities that consist of a legislative assembly (Landsting), and an administrative authority (Landsstyre) elected for a 4-year term. The government of Greenland assumed the responsibility for local matters such as conservation and environmental protection, local planning, trade and competition, welfare and health services, education, housing, internal transport, fishing and hunting, taxation, and religious communities. Additionally, the exploitation of natural resources was recognized as a concurrent power (Greenland Home Rule Act, 1978). The first election to the Greenlandic Parliament was held on April 4, 1979 and the left-wing Simut Party formed the first Greenlandic government. Undoubtedly, the Home Rule Act increased Greenlanders control over home affairs and established grounds for further expansion of power. Greenlandic Home Rule was a political reform recognizing a politically, geographically, and demographically undivided Greenland (Dahl, 1986, p. 323). One of the major autonomous decisions of the new government was a withdrawal from

7 From 1883 to 1998, around 150,000 indigenous were separated from their families and forced to attend church-run boarding school. They were discriminated and abused. They call themselves a lost generation.

the European Union which was completed in 1985.[8] At the end of the twentieth century, the Commission of Self-Governance was established to reevaluate Greenland's position within the Danish Realm. One of the major debated issues was a reduction of economic dependency on Denmark by expanding the power of Greenlandic government over natural resources. Following the Commission recommendation, a Greenlandic-Danish Self-Government Commission on Greenland's self-governance was established in 2004. Three years later, the Commission, composed of representatives appointed by the Danish and Greenlandic governments, respectively, the academia, and external experts, drafted an Act on Greenland Self-Government that greatly expanded authority of the Greenlandic government. At the same time, the US Geological Survey published a report that indicated the potential for undiscovered, technically recoverable oil and gas resources in the East Greenland Rift Basins Province (USGS, 2007). This was great news for Greenlanders who overwhelmingly voted in favor to adopt this Act. A year later, the Act on Self-Government entered into force, which gave Greenland significant autonomy. Firstly, the Act recognized Greenlanders as a people under international law; therefore, they were entitled to be represented at Danish missions. Secondly, it granted power over its mineral resources to Greenland. Thirdly, it opened the door for Greenland's independence under conditions of approval by referendum in Greenland and by the Danish Parliament. The Act authorized Greenland to secure an additional 33 areas of responsibility, divided into two categories. The first category listed areas that will be transferred to Greenlandic authorities at a fixed time, while the second category listed areas to be transferred after the process of negotiation with the Danish administration (Act, 2009). The Act also established *kalaallisut* as the only official language of Greenland, thus supporting not only the local language but also national identity. Without status, the Danish language has been substituted with English (Gad, 2017). Consequently, this new law provided for a new institutional framework leaving the Danish government with certain control over foreign affairs, security, and defense but with the Greenlandic government's consent. The Greenlandic government may sign an international agreement with a foreign state or an international organization on Greenlandic subject matter, but the Danish government should be notified. It seems that the Act on Greenland Self-Government might lead Greenland to full independence which is supported by a solid

8 When in 1973, Denmark accessed the European Community (EU), only 30% of Greenlanders supported the entry due to a dispute over the Common Fisheries Policy (CFP). In a 1982 referendum, 52% Greenlanders voted to leave the European Community. After the withdrawal, Greenland was awarded Overseas Country and Territory status for all areas except fishing industry.

majority of Greenlanders. According to HS Analyze poll, carried in December 2016, 64% of respondents found the island's independence to be of "great importance" or of "importance," while it was "insignificant" for 24% of respondents and 12% had no opinion on this matter (McGwin, 2016). As Kim Kielsen, Greenland Premier, emphasized, "We are working toward independence, but it is not a goal in itself that we should end up isolated as if we sat in the glass." He also indicated that Greenland cooperation with Denmark on defense, foreign affairs could continue (Breum, 2020).

Transition from the territory to a State brought significant changes to governance and land ownership in Alaska. American states have all the powers that are not delegated by the Constitution to the federal government, so each state has sovereignty over its territory and domestic affairs. It means that the State of Alaska is responsible, among others, for commerce, community, and economic development, environmental conservation, natural resources, labor, and workforce development. Land administration in Alaska is divided into boroughs, as opposed to counties, with the largest being aptly named "Unorganized Borough" as it has no local administration and is governed directly by the state instead. Today, around 734,000 people reside in Alaska with the Indigenous people comprising just under percent of the state's total population, which is the highest share of indigenous Americans of any state in the United States (World Population Review, 2020). Some 60% of Alaska's land is administered directly by the federal government, under the auspices of organizations such as the Bureau of Land Management and the US Fish and Wildlife Service. The federal government granted Alaskan government 28% of its total area (approximately 105 million acre). Native lands are private, held by regional and village Native corporations. They own around 12% of Alaska's land (44 million acres). Less than 1% of Alaska's land is privately owned (Government of Alaska, 2000). Alaska's economy has always depended on natural resources. Before Alaska became a state in 1959, the fishing and mining industries prevailed, although also the military has played an important part in Alaska's history and economy. During WWII, thousands of troops were sent to Alaska. The army built the Alaska Highway and many other roads, airfields, and military bases that shaped Alaska's transportation infrastructure. An oil field discovery at Prudhoe Bay on the Arctic coast in the late 1960s and the construction of almost 1,300 kilometers of a pipeline to Valdez on Prince William Sound fundamentally transformed the state. The new industry had a profound impact on economy, environment, and social structure. On the one hand, millions of dollars in royalty and tax money were injected into the Alaska economy; and on the other hand, the industry posed a great risk to the environment and altered the life of Indigenous people. Since the oil discovery, that industry has been extremely important to Alaska's financial system. Along with federal spending, Alaska's economy has been driven by the extraction industry. Although only a few thousand people have been directly employed by oil companies, many work in oil-related jobs such as oilfield service activities, oilfield construction projects, and pipeline operations. The State of Alaska receives oil royalties and severance taxes. Oil accounts for about 85% of the "unrestricted" revenue available to the State for spending for general purposes. Thus, the oil industry pays for a very large share of state government and local government employment (Knapp, 2012, pp. 18–19). The second major industry in Alaska has been seafood industry. Different kinds of fish are processed into various kinds of products. Additionally, mining industry has been also operating, although high costs of extraction and transportation as well as volatile prices make it difficult for mining projects to compete in the world markets. The oil spill of Exxon Valdez in 1989 drew international attention to environmental protection. The oil tanker collided with a reef in the Prince William Sound pouring almost 11 billion gallons of North Slope crude into the Gulf of Alaska. It was one of the largest marine oil spills in the US history (Raspotnik & Uljua, 2018). Thus, Alaska has been irretrievably altered and never the same again. Environmental concerns severely impact discussions on the future of Alaska's petroleum industry.

Indigenous Self-governments

The growth of self-governments of indigenous people in the North American Arctic in the late 1970s is related to the general recognition of the rights of Indigenous people. This development was driven to a certain extent by different Native groups in the region that look not only for the recognition of the ownership but also for reallocation of lands and resources with the management potential. The outcome of this development varies between Canada, Greenland, and Alaskan US. As Jens Dahl argued, the different kinds of self-governments reflect the balance of power between the native population and the nation-states (Dahl, 1986, p. 316).

In Canada, the right of Indigenous people to self-government is based on a unique legal constitutional framework.[9] There is a general agreement that Indigenous people's rights are inherent (Slattery, 1983, p. 232). Additionally, the Constitution Act, 1982, sec. 35, recognized and affirmed the existing

9 Section 35 of the Constitution Act, 1982 states that Aboriginal peoples of Canada include the Indian, Inuit, and Metis people of Canada.

aboriginal and treaty rights of aboriginal peoples of Canada.[10] Thus the Canadian government acknowledged the inherent right of aboriginals to self-government (McNeil, 2007, p. 31). Subsequently, Inuit and other Indigenous peoples from the North have been trying to accommodate their local governments within the existing constitutional order. They have been participating in that process through land claims, settlements, and self-government agreements. The Comprehensive Land Claims Agreements have opened prospects for self-governance that allows the Natives to control local economies and be a player in resource development activities. The agreements mandated joint territorial governments and aboriginals' boards manage wildlife and environmental protection, land use planning, and budgets and push for restoration of autonomy over their traditional land and resources. Since 1973, the Canadian government has signed over 20 comprehensive land claim agreements with the Indigenous peoples from the North.[11]

The first land claims agreements in the North were settled in Northern Quebec. The James Bay and Northern Quebec Agreement and Northeastern Quebec Agreement were signed in the 1970s. They reserved not only 14,000 square kilometers exclusively for the use of native Quebecers but also additional lands for specific hunting and harvesting. The accords also established a system of environmental evaluation for new development projects and provided the northerners with direct financial compensation for development.

The first agreement signed in the NWT created Inuvialuit Settlement Region in western Arctic in 1984. Under it, the Inuit community agreed to give up their exclusive use of the ancestral lands in exchange for certain rights such as land, wildlife management, and money. Ten years later, Inuvialuit commenced the self-government negotiations with the Government of Canada and the NWT. The three parties signed a Self-Government Agreement-in-Principle in 2015 (AIP) that prepared the ground for negotiation a final agreement on self-government (AIP, 2015).

The indigenous community in Yukon that consists of 14 small Indian bands jointly negotiated the final agreement, the so-called Umbrella Final Agreement (UFA) in 1993, and they also settled most of the claims.

10 It means that indigenous treaties before and after the Confederation are recognized.

11 In Quebec: James Bay and Northern Quebec Agreement (1975); Northeastern Quebec Agreement (1978); Nunavik Inuit Land Claims Agreement (2007); in NWT: Inuvialuit Agreement (1984); Gwich'in Agreement (1992); Sahtu Dene and Métis Agreement (1994); Tlicho Agreement (2005; Déline – Sahtu Dene and Métis – Self-Government Agreement (2016); in Nunavut: Nunavut Land Claims Agreement (1993); in Newfoundland and Labrador: Labrador Inuit Agreement (2005); in Yukon: the Umbrella Final Agreement 1993 that provides common template for agreements negotiation with 14 small Indian bands.

As aforementioned, the NCLA represents the largest Aboriginal land claims settlement in Canada. The NCLA vested Innuit with ownership of approximately 356,000 square kilometers of land (15% of Nunavut) and provided cash payments of $1.148 billion from the federal government over a 14-year period and a share of federal royalties from Crown land and natural resources. The Innuit also received the rights to use lands and resources for wildlife harvest and water and the right to participate in decision-making concerning the use management and conservation of land, water, and resources. In exchange, the Inuit agreed to cede, release, and surrender to the Crown of Canada all their claims, rights, titles, and interests to land and waters in the Nunavut settlement area (NLCA, 2.7. 1). The Nunavut Tunngavik Incorporated (NTI), the successor of the Tunngavik Federation of Nunavut that negotiated and signed the NCLA on behalf of Innuit, oversees implementation of the agreement. The agreement was a great political accomplishment, but the implementation is a continuing process. Differences in interpretations of obligations, disputes over funding, and a limited collaboration between the main partners, the Government of Canada, the Government of Nunavut, and NTI, affect that process.

In addition, the Inuit people of Labrador also contracted a land claim agreement and received the right to self-government. As a result of the act, the Nunatsiavut Government was established, representing around 6,500 Inuit from the whole province. Each of the agreement is unique but usually includes land ownership, financing, wildlife harvesting rights, conservation of land water and resources, and general management. There are also measures to support economic growth by allowing setting up resource development corporations. Some of the indigenous groups operate their own corporations in forestry, road construction, housing corporations, oil field maintenance, agriculture, and fisheries. The various Arctic people's organizations have managed to be a prominent player in a variety of comanagement arrangements designed to share decision-making also in the international bodies.

In Greenland, the process of devolution (explained above) was primarily related to the Indigenous peoples of Greenland who compose 89% of the population. Initially, through the adoption of Home Rule Act and then the Act of Self-Government, the Inuit reached the goal of democratic self-rule. Owing to that, Greenland has become a symbol of a new world democratic unity with the old world, as Eben Hopson declared at the inaugural meeting of the Greenland Legislature (Dingman, 2014, p. 2).[12]

---

12 Eben Hopson was a renowned American politician from Alaska, founder of the Inuit Circumpolar Council.

As in Canada and Greenland, Alaskan Indigenous peoples also, but to a lesser extent, influence the governance in the North American Arctic. Divided into five major groups, Aleuts, Northern Eskimos (Inupiat), Southern Eskimos (Yuit), Interior Indians (Athabascans), and Southeast Coastal Indians (Tlingit and Haida), they make up most of the population in the distant western and northern regions of Alaska. The existence of Alaska's Indigenous peoples was vastly affected by the industrialization process of the state. As the Canadian Natives, they did not fit into the rising new, white society. Jobless, harassed, and assaulted were forced to send their children to residential schools to "adjust Alaska Native students for an urban technological society, thus accelerating the breakdown of the old life" (Agukkasuk, 2012). Nevertheless, they survived adapting to new challenges. The Alaska Native Claims Settlement Act (ANSCA) of 1971 institutionalized the position of Alaskan Natives. It granted to village and native corporations, created under the act, roughly 44 million acres and $962.5 million in compensation (a half over 11 years and another half from royalty revenues) for extinguishment of Alaska Native claims to additional lands based on aboriginal title.[13] The land was divided among 220 villages and 12 regional corporations. While the communities held only surface rights to their lands, the native-run corporations would hold full title and subsurface rights and be responsible for their administration and for cash payments received from the federal and state governments. Natives were expected to become shareholders in the corporations and receive annual dividends from profits.

> *The Alaskan settlement was considered historic, not only for the speed at which it was negotiated, but for provision for subsurface rights to a portion of the land award, a first in North America.*
> *(Grant, 2010, p. 358)*

The land settlement and the establishment of native' corporations led to native self-governments formation. Indigenous people use them to pressure both the Alaska state and federal administration into accommodating their demands for hunting rights as well as for environmental protection. On the other hand, many Natives view ANCSA as an act designed to assimilate Alaskan Natives into the mainstream of American society and divested them of their land culture and tribal sovereignty (Ward Ford & Rude, 1999, p. 489). Nevertheless, pressure for full recognition of aboriginal rights of Alaskan Indians and Eskimos resulted in the 1999 State Administrative Order No. 186 that recognized the

13 A Native, according to the act, was defined as a US citizen who had at least a quarter Indian, Aleutian, or Eskimo ancestry.

right to a tribal government in what was loosely described as "Indian territory." It seems to be true that, as a result, their land claims settlement, which originally had been an attempt to assimilate Alaskan Natives, has become a "permanent institution of Native empowerment" (Grant, 2010, p. 416).

Today, federally recognized tribes have sovereign power which includes the power to form tribal governments. Most of all 299 tribes in Alaska have councils and governing bodies. They deliver various services to their members comprising health care, social services, and environmental protection and even judicial services. Additionally, the Alaskan Natives are part of four international indigenous councils in the North American Arctic that voice Indigenous peoples' concerns.

STATES OF NORTH AMERICAN ARCTIC: STRATEGIES AND POLICIES

Canadian Strategies and Policies

The Canadian approach to the North has evolved since the inception of Canada in 1867. In the first part of the twentieth century, the federal government paid little attention to the region, although Canada, to claim an unexplored land, introduced the Sector Principle or the Sector Theory in 1907 to draw a boundary. Then the North gradually gained importance during World War II and the Cold War for military and surveillance purposes (Coates, 1985, pp. 191–198). In the late 1950s and in the early 1960s, Prime Minister John Diefenbaker (1957–1963) committed his administration to secure the Canadian position in the North by economic development and military activities. Some projects were implemented, but due to high costs, the northern development was put on hold. Subsequently, Prime Minister Pierre Trudeau (1968–1979; 1980–1984) made some efforts to enhance Canadian sovereignty and to promote economic development in the Arctic, which brought about some friction with the United States. This policy was continued by Prime Minister Jean Chrétien (1993–2003) who also advocated for international cooperation between the Arctic countries in the 1990s. Current policy is based on Prime Minister Stephen Harper's (2006–2015) legacy and Liberal's agenda formulated by the government of Justin Trudeau who became Prime Minister in 2015.

Since the beginning of the twenty-first century, the North has become an important part of Canadian federal politics. Especially Stephen Harper, a Conservative, put the North high on his agenda. Harper brought back the

Progressive Conservative plan formulated by John Diefenbaker, but in different political and environmental circumstances. At least three major factors influenced the promotion of the North by the Harper government. First of all, Harper looked for ideas that would unite Canadians and increase support for his party in the future. The North, vast, with hidden wealth, perfectly fits into that picture. Secondly, from the very beginning, Harper's Arctic policy was based on the fear of external threats to Canadian sovereignty and security. To a certain extent, this idea was based on the Conservative agenda dating back to the late 1950s. Thirdly, growing international attention to the North increased what was regarded as a potential threat to Canadian sovereignty. Apparently, a combination of these three factors influenced Harper's approach to the North as a symbol of a positive nationalism and proud heritage that needs to be protected and developed (Wrońska, 2018, p. 255).

Harper declared Arctic sovereignty a key objective of his policy and strongly asserted that the Arctic is fundamental to Canada. Already in the 2007 Speech from the Throne, he had addressed specific issues relating to the Arctic such as a commitment to complete mapping of Canada's Arctic seabed, the creation of a world-class Arctic research station, and the improvement of living conditions for residents of the North. This was reflected in official statements and documents on the North, where the issue of protecting Canadian sovereignty in the Arctic was constantly highlighted, as well in symbolic gestures such as renaming Northwest Passage to the Canadian Northwest Passage in 2009. Consequently, the Conservative government launched a *Northern Strategy* that designed a new approach toward the Arctic, which focused on four pillars: exercising Arctic sovereignty; promoting social and economic development; protecting the environment; and improving and devolving Northern governance (Government of Canada, 2009).

The first and the most important priority of the *Northern Strategy* was a confirmation that Canada exercises sovereignty over Arctic lands and waters – sovereignty that is long-standing, well-established, and based on historic title, international law, and the presence of Inuit and other Indigenous peoples for thousands of years. Thus, in order to secure and enhance Canada's position in the North, the government announced several flagship initiatives. The most noticeable was an annual Prime Minister's northern tour or pilgrimage, as journalists called it. Harper made several official trips to the North, highlighting that the North is fundamental to Canadian heritage and a key to prosperity (Wrońska, 2012, pp. 109–114). These visits were also linked with the Canadian Armed Forces (CAF) Operation NANOOK, annual High Arctic sovereignty exercise that commenced in 2007. Since then, Operation NANOOK has been the CAF' signature northern operation, comprising a series of comprehensive, joint, and international activities designed to exercise the defense of Canada and security in the region. Importantly, the regional partners, the United States and Demark, have been taking part in these military exercises, thus reflecting necessity of cooperation in the harsh environment. In the Operation NANOOK-TUUGAALIK 2020, the Canadian forces trained together with the US, Danish, and French forces.

Although the Conservative government strongly emphasized sovereignty protection in its strategic vision, in reality, security in the region was not significantly improved. The widely publicized announcements of the procurement of an icebreaker and eight Arctic/Offshore Patrol Ships, as symbols to safeguard the North, were reduced and delayed. The new CCGS John G. Diefenbaker icebreaker was supposed to be operational in 2017, but the construction has not even commenced. Meantime, an old CCGS Louis S. St. Laurent was significantly refurbished to continue research in the Arctic, which means that the new one will not join the fleet soon.[14] The construction of patrol ships has been also greatly delayed and the initial number was reduced. So far, only one vessel commenced the service. Delay and scaling back also affected the naval facility in Nanisivik (Baffin Island, Nunavut), which was intended to provide fuel to the patrol ships and other vessels. The facility was expected to be operational by 2012 as a full season refueling station, but it is still under construction, only as a seasonal facility. Apart from the military exercises, the only project that was delivered to enhance security in the region was the scaled-back Arctic Training Facility in Resolute Bay, Nunavut. Opened in 2013, the Center also assists the Canadian Rangers and provides emergency response and disaster assistance to civilian agencies. The Canadian Rangers is a special, reserve force of the CAF responsible for providing a military presence and surveillance in the North. This unit collects data, reports unusual activities, and supports military operations in the region. Although the number of Canadian Rangers expanded to 5,000 from 4,000, this force is too small and ill-equipped to monitor the entire North (Wrońska, 2018, pp. 258–260). Nevertheless, the Canadian Rangers even with the limited capacity are not only the "eyes, ears, and voice" of the CAF in the North due to their familiarity with the local cultures and languages but also the voice of the military in their communities (Lackenbauer, 2013, pp. 6–7). There is a general agreement that

---

14 Canada's existing Polar Icebreaker, the CCGS Louis S. St. Laurent, the only ship in the Canadian fleet capable of year-round operations in the Arctic, is in its sixth decade of service. A replacement is needed by 2029 to support year-round marine trade, to resupply Canada's Arctic communities and industries, and to conduct search and rescue and environmental response operations in the Far North.

the growing accessibility of the Canadian Arctic requires a modernization and expansion of the Canadian Rangers (Lackenbauer & Kikkert, 2020, p. 133).

During Harper's term, there were also implemented projects other than sovereignty promotion. For example, Geo-Mapping for Energy and Minerals (GEM) project that significantly increased geoscientific data about Canada's North and a new world-class research station, the Canadian High Arctic Research Station (CHARS), at Cambridge Bay, Nunavut, started to operate in 2018. To a certain extent, all of these initiatives were supposed to support Canada's intention to claim an extended continental shelf in accordance with the United Nations Convention on the Law of the Sea that might give Canada sovereign rights over the resources of the seabed and subsoil.[15] The recognition of the claim does not depend upon occupation or express proclamation. Based on geological evidence, the UN Commission on the Limits of the Continental Shelf (CLCS) will determine the Canadian limits (UNCLOS, 1982). In December 2013, the Government of Canada delivered partial submission to the Commission and in May 2019 an additional part (Government of Canada, 2019a). It will take years for the submission to be reviewed. As of 2020, Canada's submission is 84 on the CLCS' list (UN, 2020). It is important to note that the Canadian claims depend on delimitation with the Kingdom of Denmark, the Russian Federation, and the United States. Several branches of the federal government with the assistance of territorial government and various indigenous groups worked on Canada's submission that commenced in 2006 and was concluded in 2016. Collecting data in the Arctic also required international cooperation with the United States, Denmark, Sweden, and Germany. Until 2013, the cooperation went smoothly, but in December, just before Canada was ready to make its claim, the Canadian Prime Minister Harper decided to file only a partial submission. The government decided to complete the research in order to claim the North Pole. This decision broke the trust among the countries, especially between Denmark and Canada. As Riddell pointed out, this unilateral action contravened customary international legal

15 Arctic States have rights to areas on their extended continental shelves beyond their exclusive economic zones. The United Nations Convention on the Law of the Sea (UNCLOS) explicitly recognizes the rights of coastal states such as Canada over the natural resources of the seabed and subsoil beyond 200 nautical miles from their coastal baselines and sets out a process by which a state may determine the limits within which it may exercise those rights. Under Article 76 of UNCLOS, the coastal states may file a submission to the UN Commission on the Limits of the Continental Shelf. According to UNCLOS, the state has to provide evidence that its continental shelf extends beyond 200 nautical miles. Annex 1 of the rules of procedures states that matters regarding disputes which may rise in connection with the establishment of the outer limits of the continental shelf rest with States.

practice and the use of equidistance principle and undermined cooperation with its neighbors (Riddell-Dixon, 2017, p. 239).

When Trudeau became the Prime Minister, the Arctic policy was revised, although several flagship initiatives commenced under the 2009 Strategy were continued. The new government withdrew from concentrating on sovereignty protection to focus on Indigenous people, environmental protection, and multilateral cooperation. This departure reflected also the commitment to codevelop the Northern policy with Indigenous people (Kikkert & Lackenbauer, 2019, p. 3). Several federal, provincial, and local institutions took part in drafting this policy. Launched on September 10, 2019, Canada's Arctic and Northern Policy Framework (ANPF) comprises primary goals and ambitious long-term objectives to guide the Canadian government's activities in the Arctic to 2030. Additionally, the framework reflects to a greater extent interests and visions of Indigenous peoples, territorial, and provincial partners. The Inuit authored the part of goals and objectives that should be implemented in their community. Territorial and provincial governments outlined the needs in their domains such as economic development, infrastructure, health care, and postsecondary education (Government of Canada, 2019b).

When we compare Harper's and Trudeau's strategies, the priorities are mostly the same, just the emphasis differs. Harper put the major accent on enhancing sovereignty in the Arctic. Trudeau lists area of health, infrastructure, and economic development of the Northern communities as a top priority by stressing the need for strong sustainable, diversified, and inclusive local and regional economies (Government of Canada, 2019b). Both strategies outlined the impacts of climate change on security, environment, nature, and social life and promoted social and economic development, environmental protection, and improving Northern governance, but ANPF pledges to better align Canada's current national and international policy objectives with the people of the North by continuing to support security and stability in the region. Part of a plan is to support for Search and Rescue reaction and responsiveness for Arctic visitors and residents by enhancing the CAF' presence in the region. The ANPF prioritized international cooperation in the Arctic, which was in the background of the Harper Strategy. The Arctic Council was seen as a major forum, but bilateral dialogues, even with a call to restart regular dialogue with Russia, were also predicted. Canada has not been an advocate of formalizing regional cooperation among the Arctic Ocean coastal states, but has largely preferred bilateral arrangements (Koivurova, Molenaar, & Vanderzwaag, 2009, p. 274).

Canadian government also confirmed its strong involvement in promoting the sustainable development, as it was set in the United Nations 2030 Agenda for Sustainable Development (Government of Canada, 2019b). Judging from the way the Liberal government was handling initiatives commenced by the Conservatives, it was apparent that Canada will continue to demonstrate sovereignty in the Arctic. The ANPF chapter of Safety, Security, and Defense pledged for increase of the military presence and improving emergency awareness, especially concerning marine safety.

The ANPF primarily focused on northern communities and international cooperation in conserving the Arctic environment. It clearly indicates that decisions about the Arctic will be made in partnership with the northerners and the development should be sustainable, holistic, integrative, cultural, economic, and environmental considerations. It provides ambitious goals and several principles but to a great extent omits the implementation measures and financing. It also lacks consistency; in some instances, the framework is very specific and in others very general. As critics point out, the ANPF has no budget, timelines, or clear plan on how to address the identified challenges. It is a half-baked chaotic mess that lists well-known issues without any plan for action (Exner-Pirot, 2019; Greer, 2019; Kikkert & Lackenbauer, 2019).

Nevertheless, this framework will guide the Canadian policy in the Arctic in the near future. Already, in order to implement some of the ideas, the 2019 Canadian federal budget allocated more than $700 million over 10 years to the northern communities. The specific measures include transforming Yukon College into a hybrid university, support for Dechinta Centre for Research and Learning in the NWT and Nutrition North Canada program, institutional help to fight addictions, investments in clean energy projects such as expansion of Taltson hydroelectricity and in access to Internet, financing the Polar Continental Shelf Program, ongoing research in the Arctic, and the project of marine conservation area in the High Arctic Basin (Government of Canada, 2019b).

Danish Strategies and Policies

Thanks to the geography of Greenland, Denmark is an Arctic country but Greenland has been striving for full independence since the late 1990s. There is no doubt the Home Rule Government in Greenland institutionalized the claims for sovereignty, as expressed by the leading political parties. It seems that sooner or later Greenland will become an independent country, but so far, the Danish politics and strategies have been focusing on equal partnership, thus trying to keep up the institutional ties with Greenland.

As already indicated, the Danish strategies toward Greenland have evolved from colonial relations to full partnership as affirmed in the Act on Greenland Self-Government. For Denmark, sovereignty over Greenland, even limited, means that Denmark is and will remain an Arctic country. Announced in 2011, the Kingdom of Denmark Strategy for the Arctic 2011–2020 (Strategy) had to take into account the arrangements of the new self-government in Greenland (Kingdom of Denmark, 2011). It was prepared and signed not only by the Prime Minister of Denmark but also by the heads of Greenlandic and Faroese governments. This Strategy encompassed the opportunities and challenges set out by the three parts of the Realm, with the aim to promote information exchange and coordination of efforts related to the Arctic.[16] Thus, it provides general framework of principles and activities of the Realm in the region. The Strategy was grounded on four pillars. The first pillar concentrated on peace and security but with an emphasis on sovereignty protection. The second pillar promoted self-sufficient growth and societal sustainability. The third one focused on respect for the Arctic fragile environment and the last one on close cooperation with the international partners. Nevertheless, as it was emphasized, this Strategy is designed to bring a development that benefits the inhabitants of the Arctic, while respecting their right to resources and culture preservation.

Primarily, the Strategy highlighted the value of cooperation among the five Arctic States within the frame of international law which provides the basis for protecting peace and security in the region. So, it promoted joint activities over continental shelf claim. As in Canada's strategy, special attention was given to the CLCS submission. The Danish government confirmed its declaration to file the full submission by December 2014 with the exemplary cooperation, as highlighted, with other countries including Canada. Overall, Denmark with the government of Greenland filed the last submission in December 2014 to claim around 90,000 square kilometers of the continental shelf. In 2016, an official delegation from Greenland and Denmark presented claims to the CLCS in New York (MFAD, 2016) but the assessment process may take several years. Additionally, Danish submission overlaps with Canada's, Russia's, and Norway's claims. According to UNCLOS, the States with overlapping claims have to negotiate separate delimitation agreements, so the delimitation process will be further prolonged. The Danish government pledged to act in accordance with international law and work for peaceful cooperation in accordance with the foundations established by the Ilulissat Declaration.[17] Secondly, the attention was given to maritime safety due to higher maritime traffic, including cruise

16 The Kingdom of Denmark is constitutionally a unitary state but with two autonomous territories: the Faroe Island and Greenland. This forms the Danish Realm (Danmarks Rige).

17 Ilulissat Declaration of 2008 was signed by five Arctic coastal States. The States committed to resolve disputes and overlapping claims through negotiation.

ships around Greenland. The Danish government pledged to improve surveillance by Port State Control (PSC) of cruise planning, GRENPOS reporting system or Long-Range Identification System; expand infrastructure; update nautical charts; and implement the Polar Code. During the last 10 years, new measures on safe navigation in Greenland waters were implemented. Two ship reporting systems have been established: GREENPOS, for ships engaged on voyages to and from Greenland ports and places of call, and KYSTKONTROL, for ships engaged in costal trade; additionally, order on pilotage around Greenland was implemented. The Danish Navy has been updating most of the seabed mapping, and the Danish Geodata Agency has been working on electronic navigational charts. The Polar Code entered into force in 2017 (DMA, 2020). The opening of Nuuk's new port facility Sikuki Harbor in 2017 was part of the infrastructure investments, as promised in the Strategy. Finally, the last point focused on exercise of sovereignty and surveillance. It proposed four overriding initiatives such as forming a joint Arctic Command, the establishment of an Arctic Response Force, a risk analysis of the maritime environment around Greenland, and an analysis of the potential of international cooperation in the Arctic. Most of the initiatives were fulfilled. The Arctic Command was established on October 31, 2012. The Arctic Command was reinforced by the Danish Arctic Response Force that uses existing military capabilities adapted for Arctic operations. The Marine Environmental Risk Assessment Report concerning Greenland and the Arctic was delivered in 2015. The analyses presented in this report quantify and describe the likelihood of marine accidents with and without pollution by spill of oil products, oil/chemical cargo products, and fuels, as well as corresponding spill volumes. The analysis also includes prognoses on future ship traffic around Greenland until 2027 (DNV GL, 2015). The issue of international cooperation has been analyzed and discussed on several occasions. The shape of it is still being developed.

The second and the third pillars of the Strategy are intertwined. Self-sustaining growth and development depend on respect for the Arctic's vulnerable climate, environment, and nature. Taking into consideration the need to exploit natural resources, the Strategy created a list of principles that should be followed in the future. According to them, mineral resources shall be exploited under the highest international standards of safety, health, environment protection, preparedness and response, and transparency with high returns for society; living resources, including fish, shellfish, and marine mammals, shall be harvested in a sustainable manner based on sound science; new opportunities in the Arctic must be exploited in close cooperation with industry, and an optimal regulatory framework will be created for exports and investments;

research and training efforts must support the development of industry and society in the Arctic; and finally, the use of renewable energy sources must be increased significantly (Kingdom of Denmark, 2011, p. 23). As in the Canadian strategies, much attention has been paid to the consequences of climate change and its impact on the Arctic. It proposed to reinforce research monitoring and preventive steps against marine pollution.

The last part of the Strategy calls for global cooperation on the Arctic, including focus on climate change, protection of nature and environment, endorsement of strict global maritime rules, and protection of Indigenous people's rights not only within the frame of the Arctic Council but also globally. It also underlined the value of bilateral cooperation with Canada, the United States, Norway, and Iceland.

The Strategy set a stage for the Realm's activities in the Arctic until 2020. To a different extent, most of the initiatives and policies have been implemented. The new strategy for years to come is being developed but delayed by COVID-19. It should be ready by early 2021(Bye, 2020).

The future relationship between Greenland and Denmark has been lately a central issue, especially with the growth of global focus on the Arctic. The main question that the government of Denmark has to approach is how long will Denmark belong to the group of five coastal Arctic states? Since the beginning of the twenty-first century, the Danish government tried to democratize the self-governance in Greenland but within the Danish borders. After the Self-Government Act, the Danish authorities promoted close cooperation not only underlining the economic benefits but also acknowledging the unique identity of Greenlanders. Nevertheless, the self-government institutions strengthened pro-independence movements. As Adam Grydehøj claims, Greenland's colonial experience has trapped the desire for independence in a Denmark-oriented conceptualization of Greenlandic identity, which prevents the pursuit of potential political and economic future (Grydehøj, 2016, p. 102).

There is a consensus among Greenlandic political parties and society that Greenland should became an independent country, which the Danish government is aware of. Although today it looks as if it is a rather long-term aim. Both parts are interconnected. Greenland is the main trade partner of Denmark, an importer of Danish labor force, and secures Danish position as an Arctic player. Denmark provides subsidies and guarantees peace and security in the region. A full independence while dependent on Danish grant would be a challenge even if the mining industry starts to operate on a larger scale. This a time- and money-consuming industry before it brings the profits. According to the report *To the Benefit of Greenland*, it would require 24 large-scale mines operating simultaneously in order to zero out the block grant (Minik, 2014).

It also would be very hard for Greenland to respond to external security concerns. As Mikkel R. Olsen argues,

> *...the Realm's ability to manage external challenges is in this relation directly linked with its ability to appear as one unified actor. If not, they risk being played out against each other by the superpowers.*
>
> *(Bye, 2020)*

It means that close dialogue and cooperation on equal terms within the Realm is crucial, especially with more States operating in the region. This direction was confirmed by Jeppe Kofod, the Danish Foreign Minister. For him, close dialogue concerning foreign and security policy in the entire modern and more generous Realm is of vital importance. The current Danish government formed in 2019 by Social Democrats, led by Mette Frederiksen, is more open to the idea of new arrangement with Greenland than its predecessor.

It seems that both sides should reevaluate their relations. As Jacobsen rightly pointed out, it should be an essential part of Denmark's strategy to improve the current relations with Greenland. This can be done through a more open, honest, and equal dialogue; more mutual knowledge exchange; by emphasizing and encouraging common values and relationships; and by giving higher economic priority to the Arctic (Jacobsen, 2016). One of the possible solutions is a federal political system which provides for the combination of shared rule and self-rule. The powers of the State would be divided between two levels of governments of equal status, not subordinate to each other within their own areas of jurisdiction. In the Danish Realm scenario, the security, defense, international politics, and currency could belong to the Realm governments while the rest to autonomous units.

American Strategies and Policies

The United States is an Arctic state by virtue of its purchase of Alaska from Russia in 1867. Alaska which has both territory above the Arctic Circle and a coastline on the Arctic Ocean makes the United States an Arctic country. Throughout its history, the United States has seldom recognized this fact, although it produced such outstanding polar explorers like Robert Peary and Richard Byrd, as well as modern-day Arctic scientists such as Robert Corell. In addition, the Arctic was central to the United States' nuclear deterrent posture during the Cold War. Nevertheless, the Arctic has seldom figured prominently in US policy discussions. "Thus the United States may be characterized as the 'reluctant' Arctic power" (Huebert, 2009, p. 2). The alleged "reluctance" of the United States is a matter of opinion, expressed with particular ease especially if you are a university professor and a Canadian with a typical love exhibited by close neighbors of unequal stature. However, there may be more than a grain of truth in this statement, as echoed by many Americans themselves.

"Unfortunately, the US national-level focus on Arctic issues and policy is quite modest, a factor that will undermine and limit US capabilities as leader of the Arctic Council" (Rosenberg, Titley, & Wiker, 2014, p. 2). Although this organization is not the only platform to influence Arctic policy and activity, it is an important one and deserving increased US attention. Despite the appointment of the former US Coast Guard Commandant Admiral Robert J. Papp Jr as a special representative for the Arctic and a number of released national roadmaps and strategies (Navy Task, 2014; The White House, 2013; US Coast Guard, 2013), the Arctic remains a policy and investment afterthought (US House, 2015; US Senate, 2015).[18]

Some experts claim that although Alaska is an important player in Arctic affairs, the United States has a lackluster record of leadership in the region.

> *Arctic regional cooperation began in earnest in the early 1990s, following the collapse of the Soviet Union. The kind of matters the Arctic Council was given a mandate to address – environmental protection and sustainable development – are motherhood and apple pie kinds of issues that all Arctic nations could accept and support. But the United States chose to play a mostly minor role in the Arctic Council, ensuring, for example, that military issues were excluded from discussions, and that no defined monetary contributions were assessed.*
>
> *(Exner-Pirot, 2015, p. 3)*

On January 9, 2009, just before leaving office, President George W. Bush signed a new Arctic policy which acknowledged the "growing awareness that the Arctic region is both fragile and rich in resources" and ensured that the "economic development in the region [be] environmentally sustainable," that institutions promoting cooperation among the eight Arctic countries be strengthened, and that the Indigenous peoples of the Arctic be involved in

18 Admiral Robert J. Papp Jr, a distinguished officer, was appointed as US Special Representative for the Arctic in 2014. He retired from this position in 2017. The Obama administration created this position to better coordinate the American chairmanship in the Arctic Council. The Trump administration eliminated this position.

decisions affecting them (The White House, 2009). This was the first revision of US Arctic policy in 15 years. The primary focus was on national security, with the warning that the United States "is prepared to operate either independently or in conjunction with other states to safeguard [its] interests" (Grant, 2010, p. 439).

Then Obama administration formulated its official strategy on the Arctic, which was officially released in May 2013.

> *Through the National Strategy for the Arctic Region, we articulate our strategic priorities to position the United States to meet the challenges and opportunities that lie ahead. We will seek to prioritize and effectively integrate the work of Federal departments and agencies with activities that are already underway in the State of Alaska and at the international level. And we will partner with the State of Alaska and Alaska Natives, as well as the international community and the private sector, to develop innovative solutions and new ways of operating. The Arctic is changing. We must proceed, cognizant of what we must do now, and consistent with our principles and goals for the future.*
>
> *(The White House, 2013, p. 1)*

In the US Arctic Strategy, there are three major points. Firstly, the document notes a desire for the advancement of national interest with consideration to all activities (commercial, scientific, and military) in the Arctic whether by sea or air with consideration to international law. Secondly, the pursuit of responsible Arctic stewardship by encouraging environmental protection and resource conservation is clearly mentioned as a priority. Thirdly, the document emphasizes the importance of strengthening international cooperation through both bilateral and multilateral relations, including working within the Arctic Council. Additionally, the Strategy was based on three guiding principles: Safeguard Peace and Stability, Pursue Innovative Arrangements, and Consult and Coordinate with Alaska.

Before once again gaining the attention of the US government, the Arctic was a concern almost exclusively of the State of Alaska (Bergh, 2012, pp. 13–14). Alaska's lack of influence in the federal government has led to tensions between the state and federal governments, notably over oil drilling in national wildlife refuges (including the Arctic National Wildlife Refuge) and the reduction of federal grants earmarked for Alaska (Bolstad, 2010). The official Arctic policy is something quite new in American politics, as previously all that was available were Presidential Directives, and it states that

> *The United States is an Arctic Nation with broad and fundamental interests in the Arctic Region, where we seek to meet our national security needs, protect the environment, responsibly manage resources, account for indigenous communities, support scientific research, and strengthen international cooperation on a wide range of issues.*
> *(The White House, 2013, p. 2)*

Additionally, in the initial address by President B. Obama, Alaska and Alaskan Natives come first, which is understandable, and only then come international organizations. However, given the United States system of government and the relative political and economic independence of its states, even the State of Alaska already poses problems for Washington D.C. The proposed focus on climate change has opened a gap between the federal government and Alaskan stakeholders "when Alaskan legislators recommended focusing on jobs and economic opportunity, as well as practical issues such as suicide prevention and sanitation facilities" (Exner-Pirot, 2015).

The other two major documents (January 2014 Implementation Plan for National Strategy for Arctic Region and January 2015 Executive Order for Enhancing Coordination of Arctic Efforts) basically elaborate on the "Strategy." The plan outlines some 36 specific initiatives that present objectives, measures to evaluate the objectives, and the names of the federal agencies to be involved.

The November 9, 2016 Press Report about Preparation of New Arctic Policy, which appears to sum up the American efforts in terms of guiding principles for the Artic region, states the following:

> *Under the cover of the 2016 election season, the Obama administration has been busy crafting a new Arctic policy that will have significant environmental, economic, and national security ramifications. It is likely to be left incomplete, with final decisions being left for the next administration to consider in 2017.*

A few signs of its development emerged. Although it was overshadowed by electoral politics, the new US Arctic policy is worthy of public consideration, as it will likely shape the business and political climate in the region for decades to come. There are three apparent objectives in this policy. First, the government will seek to adequately protect the Arctic environment, particularly in Alaska. Second, the US will attempt to bolster the economy by taking full advantage of the region's potential for hydrocarbon development. Third, the US will re-establish itself militarily in the region by placing strategic assets in Alaska (Anderson, 2016). The sentence seems to be extremely important as it clearly shows that the implementation of the "Strategy" is basically left to the new administration. One should also mention that the United States held the chairmanship of the Artic Council from April 24, 2015 to May 11, 2017, a period which began during the Obama administration and continued into the first 16 weeks of the Trump administration.

In December 2016, the Executive Office of the President released Interagency Arctic Research Policy Committee's (IARPC) Arctic Research Plan: FY2017–2021 to advance research in areas of common interest to IARPC Agencies. The IARPC 5-year plan was drafted with contributions from all IARPC agencies and reviewed by the public. It intentionally builds on the strong intellectual accomplishments and ideas of the research community at the federal, state, local, and tribal levels as well as inclusion of ideas from the academic community, nongovernmental organizations (NGOs), and industry.

As for the attitude presented by the President of the United States, Donald Trump, toward the Paris emissions target, one can only quote a statement made by a representative of the EU that "we tried very hard to convince the President, but we have not succeeded." President Trump called his decision to pull the United States from the Paris climate agreement "a reassertion of America's sovereignty" and made a comment that he was "elected to represent the citizens of Pittsburgh, not Paris." The White House says America will follow the lengthy exit process outlined in the deal, meaning it will remain in the agreement (at least formally) for another three-and-a-half years – taking us right up to the next presidential election in November 2020.

Though the United States will remain part of the United Nations Framework Convention on Climate Change, Trump declared: "As of today, the United States will cease all implementation of the non-binding Paris accord" (Hunt, 2017).

So far, President Trump has been rather quiet on the Arctic front, with perhaps the notable exception of his greenlighting drilling in the Arctic National Wildlife Refuge (contested in court) and an announcement that the United States is planning to build six new icebreakers.

> *Congress Authorizes Six Icebreakers in Pentagon Bill. The nation's annual defense policy bill cleared Congress Wednesday with a pay raise for the troops and a provision allowing up to six icebreakers. If the icebreaker part sounds familiar, that's because last year's defense authorization bill had something similar, but it was stripped out before final passage.*
>
> *(Alaska Public Radio. Arctic Update, 2017)*

The decision to eliminate high-level diplomatic positions within the US State department, including the special envoys for climate change and the

Arctic, sent a message of downgrading the US interests in the Arctic to the international community. It showed the world that Trump turned his back on the Arctic (Enge, 2017).

In spite of various consulting bodies' calls for the United States to ratify UNCLOS, the prospects of attaining it seem to be rather bleak. President Obama was not able to do it because of the Republican dominated Congress while President Trump does not seem to be interested. However, there are indications about the current administration's attitude.

> *The United States is unlikely to accede to UNCLOS in the near term, or perhaps ever. However, this does not mean that the United States should not take every action necessary to secure oil and gas resources on its extended continental shelf in the Arctic Ocean, the Gulf of Mexico, and throughout the world. The United States can accomplish this end while acting as a sovereign nation, continuing the tradition of American Presidents in proclaiming the nation's maritime and resource rights, and without acceding to a deeply flawed treaty or seeking the approval of an international commission of experts housed at the United Nations.*
>
> *(Groves, 2012)*

In the report on the Arctic strategy presented to the US Congress in June 2019 by the Department of Defense, there was an update of the strategic objectives for the Arctic region. The three major objectives were outlined, such as building awareness, enhancing strategic operations, and strengthening the rules-based order in the Arctic (Department of Defense, 2019).

CONCLUSIONS

The allocation of more power to subnational governments significantly transformed the governance in the North American Arctic. Until the 1960s, across the North American Arctic, the subnational governments had a very limited authority, and the relations between the Indigenous peoples and the central governments were generally based on colonial principles. The Canadian Constitution of 1867 assigned no powers to territories which fell under the federal jurisdiction. Additionally, it provided the federal government authority over Indigenous peoples and their traditional territories and resources. So, the Canadian North was governed by Ottawa. Similarly, Greenland until 1953 was considered a Danish colony; then it became a part of Denmark but with a limited authority, so it was managed by Copenhagen. Equally, Alaska was administered by the federal government. Then due to decolonization, democratization, and decentralization processes, the Canadian territories were granted more responsibilities, Greenland became a county of Denmark, and Alaska gained the statehood. Devolution of land and resources and self-government agreements contributed to an increased participation of northern communities in decision-making. On the one hand, the governments have received more power, while on the other hand, the process made them more accountable and fostered governments' responsiveness, since the territorial governments are more likely to be sensitive to local concerns and contexts (Alcantara, Cameron, & Kennedy, 2012, pp. 330–331). For example, that is why the environmental assessment process is based on a comanagement approach between the local communities.

Therefore, today subnational governments function autonomously according to the democratic principles. Following the negotiation of a comprehensive land claim agreement, indigenous self-governments have been established and added additional level of authority in the region. The Inuit and other Indigenous peoples in the North American Arctic have formed various forms of self-government. In the case of Greenland, the Inuit self-government practically took over the decision-making process, similarly in Nunavut. Alaska's North Slope and Arctic Slope have borough governments with a legislative assembly and planning commissions, similar to a county within American states.

The local authorities became a major player in the North American Arctic, but they face a lot of challenges. The major one for Canadian territories and Greenland is a financial self-sufficiency. Both governments rely heavily on subsides from Ottawa and Copenhagen, respectively. Although the region contains a wealth of mineral resources, the governments have to deal with the struggle between the extraction that brings income and employment and the potential risks to the environment. They also must contend with the colonial past and solve social problems. In spite of its somewhat mythical nature, Alaska is a living organism and seems to be at the moment in a dire need of some domestic attention. Just as the Arctic, it is the state most vulnerable to climate changes with all the consequences following it. The largest American state may soon be the subject to ecological and human disaster, all due to the melting ice. The main sources of income for the state are oil- and gas-related industries, including services, and above all – the federal government. With the unpredictability of oil prices and their volatility, although Alaska provides some energy security for the United States, one must always keep in mind that oil and gas extraction is much less expensive in other parts of the country and that the prices are much more favorable even if oil and gas are imported. It is not that history has not shown us what some oil-dependent countries do in order to secure the future of their inhabitants once the oil and gas extraction era is over. No such plan seems to exist for Alaska. The promises of oil returns might turn into eroding areas, floods, changes in the environment, including flora and fauna, and the irrevocable necessity of migrating and adapting to a new way of life for many Alaskans.

"Arctic sovereignty" became a key phrase in political discourse during Stephen Harper's decade and is still very much present today. Stephen Harper put forward a very ambitious plan to protect sovereignty in the North and strongly promoted it. This logic of "defending sovereignty" from foreign challenges, as Franklyn Griffiths highlighted, brought a shift from past governments, which favored recognition – persuading others to accept Canada's claims without demonstrating a capacity to enforce them – to the Harper government, which favored enactment (Griffiths, 2009, p. 435). The government actions left the impression that the Northern territory was at risk. The small disputes with Denmark and the United States did not pose any serious threats. On the contrary, the Canadian authorities put more strain on these relationships by renaming the Northwest Passage to the Canadian Passage or deciding to claim the North Pole. The United States and other maritime countries still consider the Passage an international strait. It seems that Stephen Harper used the North as a political means to unite Canadians around threatened sovereignty in order to induce a positive sense of nationalism. These efforts were strengthened by his staging of his Northern trips and the military exercises, all of which were very well covered by the media. There is no doubt that Stephen Harper rediscovered the Arctic for Canadians, and although the Liberal Government prepared its own Arctic roadmap, the main frame is based on the predecessor's concept. Justin Trudeau's framework is full of pledges and promises regarding reducing poverty and promoting sustainable development, but it seems that it lacks the coherent practical plan on how to implement them. There are several new initiatives in the framework and ambitious plan to close the gap between people living in the North and the rest of the country, but it does not specify how to do it. What is important is that this framework confirmed that Canada restored the value of multilateralism to the Arctic.

Denmark's Strategy had to take into account the arrangements of the new self-government in Greenland. It was prepared with cooperation of Greenlandic and Faroese governments. This Strategy encompassed the opportunities and challenges set out by the three parts of the Realm. Just like Harper's, the Strategy was based on four pillars, with the special attention given to the CLCS submission. The first one concentrated on peace and security and sovereignty protection, which was also Harper's priority. The promotion of self-sufficient growth and societal sustainability, which was a leading theme of Trudeau's framework, was included in the second part. The third one focused on respect for the Arctic fragile environment. The last part of the Strategy called for global cooperation on the Arctic, including focus on climate change, protection of nature and environment, endorsement of strict global maritime rules, and protection of Indigenous people's rights not only within the frame of Arctic Council but also globally. It also underlined the value of bilateral cooperation with Canada, the United States, Norway, and Iceland. Nevertheless, the main goal of this Strategy focuses on developments that benefit the inhabitants of the Arctic, while respecting their right to resources and culture preservation.

The US Arctic Strategy of 2013 was based on or rather informed by the guiding principles of safeguarding peace and stability, making decisions by using the best available information, pursuance of innovative arrangements, and consultations and coordination with Alaska Natives. The US Arctic is built on three fundamental lines of effort, all of which are to advance US interests: security, pursuance of responsible Arctic Region stewardship, and strengthening international cooperation. The first one pledged to enable US vessels and aircraft to operate in the Arctic in order to support safe commercial and scientific operations and protect security in the region. The second one focused on responsible management in the region that included environmental protection, conservation of its resources, and better understanding of the Arctic based on research and traditional knowledge. The last line promoted international cooperation through bilateral relationships and multilateral bodies.

It seems that in the United States, there have existed two parallel trends: the first one being a genuine concern of the scientific community about the climate changes and the urging of the necessity to act (documented by thousands of documents and followed by reports to the Congress), strengthened and supported by the funds allocated federally to the research, multiple NGOs, private organizations, and simply people of good will. The first trend seems to be accompanied by the second one – the lack of action on the side of the politicians. If they ever cross their paths, perhaps something good will happen.

When we compare current Canadian, Danish, and American policies and strategies toward the Arctic, we can identify well-defined goals and objectives. All the governments acknowledge that climate change profoundly impacted the region, so they try to respond to it. All of them prioritized the issue of security, sustainable development, and international cooperation in the region, although Canada and Denmark added a well-being of indigenous population as a main concern. The international cooperation especially through the Arctic Council, which is the major forum of negotiations, complemented by the international law that applies to the Arctic Ocean, is important to all the parties.

The North American Arctic States highlight the value of regional cooperation but either within the whole Arctic region or bilaterally. It seems that the States do not pay attention to the geographical region and do not foresee it as a platform for joint activities.

3

THE NORTH AMERICAN ARCTIC GAMES: BETWEEN COOPERATION AND RIVALRY

INTRODUCTION

Contrary to appearances, it was not globalization that caused the problems, but they were its harbinger. As early as the twentieth century, attention was already drawn to the environmental problems threatening all humanity. It seemed that they should have instigated States to cooperate more closely in order to solve them. As O'Brien and Leichenko point out, such actions were the premise for more lasting coexistence and even reconciliation between the warring political blocs (O'Brien & Leichenko, 2000, pp. 221–232). An attempt at solving the problem of shrinking polar bear population initiated cooperation that was finalized by a treaty on the protection of polar bears. In 1973, this document was signed in Oslo by five Arctic States: the United States, Russia, Denmark, Norway, and Canada. The wording of it not only highlights Arctic flora and fauna, but above all the special role of polar bears in this ecosystem (Agreement on the Conservation of Polar Bears, 1973). Despite the signal for cooperation, until the end of the 1980s it is difficult to notice examples of cooperation on issues such as: nuclear threat, depletion of natural resources, threats to the human environment, structurally conditioned underdevelopment, and/or demographic trends. The reasons for the lack of success were seen in ideological antagonisms or strategic rivalry between the powers (Jonson, 1990; Ugarriza, 2009). It can be seen that until the 1990s the interest of international actors in the Arctic was military in nature and strictly connected with ensuring world security and peace (Renner, 1942, p. 61). Only Gorbachev's speech on cooperation in the Arctic gave an impulse for its development between the Arctic States. He called for cooperation in the field of scientific research, environmental protection, infrastructure development, and management of Arctic resources (Gorbachev, 1987).

ROUND ONE AND A NEW HAND: COOPERATION

The Background

The mentioned events indicate that in the first round of the new hand, the States decided to cooperate with each other. This behavior of the countries was conducive to drawing the attention of international actors to global issues such as: climate warming, protection of the Arctic environment, and protection of Indigenous peoples. As a result, a variety of governmental and nongovernmental actors began to emerge in the international arena, which took up the challenge of solving the problems occurring in the North American Arctic. Nevertheless, it should be emphasized that North American Arctic affairs were far from being managed in a uniform and unified way. This was due to the multitude of entities involved in the region and their coalition capabilities. They were realized within the framework of different cooperation structures, which Oran Young describes as a "mosaic of cooperation" (Young, 2005). It is difficult to analyze all the Arctic cooperation initiatives that emerged in the 1990s. Therefore, this chapter discusses only selected examples of cooperation between States, and governmental organizations and nongovernmental organizations (NGOs), in which all the States of the North American Arctic, i.e. Canada, the United States, and Denmark, participated. The aim of the analysis is to present an image of relations and interactions between them and toward other entities. Therefore, the analysis will focus on the most important institutions and selected projects that have contributed to the Arctic cooperation of the North American Arctic States such as: International Arctic Science Committee (IASC), Greenpeace, World Wildlife Fund (WWF), Arctic Council, and Inuit Circumpolar Council (ICC).

Attempts at finding a solution to the problems that the Arctic had to face in the early 1990s can be considered a benefit, or otherwise the gains attained by players as a result of cooperation, in line with the logic of Nash's balance (Nash, 1950). Obviously, they are neither measurable nor quantifiable; nor can they be compared to the financial benefits of exploiting Arctic resources and the losses associated with pollution in the region. However, while recognizing the scientific contribution of naturalists, who show the expected future consequences of destruction of the Earth's environment and climate warming, it should be pointed out that the benefits of long-term environmental protection are much greater (Masson-Delmotte et al., 2019) than the short-term profits from the exploitation of raw materials (Watkiss, Downing, Handley, & Butterfield, 2005).

The media newsworthiness of these topics also played an important role in building the foundations of Arctic cooperation in that period, which contributed to the increased attention paid to the Arctic by the international public opinion, the international community, and the media (Christensen, Nilsson, & Wormbs, 2013, pp. 26–51). On the one hand, it is important to understand that it is not only the Arctic States that influence the issue of global warming, i.e. the effects that naturalists observe, for example, on the melting of glaciers or changes in natural conditions resulting from global warming. On the other hand, not only the Arctic players, but also those countries that are located in a completely different geographical zone and climate feel the benefits of the environmental measures taken by the entities involved in the region. Thus, we are dealing with links that go far beyond the geographical region of the Arctic.

International Arctic Science Committee

On the basis of discussions about the climate and possibilities of cooperation in this region, projects bringing together entities initiating such activities on an international level began to be created. One of the first such initiatives was the establishment of the IASC in 1990. It is an NGO which aims at supporting scientific cooperation in the field of Arctic research. The most important functions of the IASC include: providing objective and independent scientific advice on research issues, communicating information on research, taking actions to protect, share, and access scientific research results, and supporting international access of researchers to geographical areas of the Arctic, including the sharing of resources and logistics (Anon, n.d.). In 1990, only eight Arctic States signed a document establishing the organization. It should be emphasized that the discussed Arctic States, i.e. Canada, the United States, and Denmark, and therefore also Greenland, operate in this organization (The IASC Handbook of Procedures & Guidelines, 2020).

The results of research conducted in the Arctic within the IASC are, therefore, a source of knowledge that should be the basis for the activities of other Arctic entities – including States and multinational corporations. This is not always the case. The weakness of IASC, on the other hand, is the issue of financing and transforming the research results into proper actions for the entities for which ecology is not always a priority. Therefore, in order to speak about effective governance in the Arctic, there must be coordination with the governments of the Arctic Eight and other governmental and NGOs.

What is most important for this organization, in the light of the Nash theory, is the approach to cooperation: "Rather than defining human and environmental boundaries, IASC tries to bridge those boundaries." IASC is also committed to recognizing that traditional knowledge, Indigenous knowledge, and "Western" scientific knowledge are coequal and complementary knowledge systems, all of which can and should inform the work of IASC. Overall, IASC promotes and supports leading-edge interdisciplinary research in order to foster a greater scientific understanding of the Arctic region and its role in the Earth system.

To achieve this mission, IASC initiates, coordinates, and promotes scientific activities at a circumarctic or international level, provides mechanisms and instruments to support scientific development, provides objective and independent scientific advice on issues of science in the Arctic and communicates scientific information to the public, seeks to ensure that scientific data and information from the Arctic are safeguarded, freely exchangeable, and accessible, promotes international access to all geographic areas and the sharing of knowledge, logistics, and other resources, provides for the freedom and ethical conduct of science, promotes and involves the next generation of scientists working in the Arctic, and promotes polar cooperation through interaction with relevant science organizations (About IASC, n.d.).

It is also interesting that this organization does not only focus on the study of natural processes but also addresses social and economic issues. Furthermore, it seems beneficial that there is no focus on the political or economic aspect dividing the participants. Therefore, the scientific character of this cooperation appears to be very valuable. The activities of, e.g., Calotte Academy can be mentioned,

> *...which has been arranged annually since 1991, has a participatory approach with visits, sessions and 'hearings' in several destinations in the European Arctic, across (national) borders (Finland, Norway, Russia, Sweden, and Sapmi), with local audiences and expertise. It aims to bring together academics and other experts, policy-shapers and – makers to discuss, debate and brainstorm on relevant global, regional and local issues with an intention of sharing information and ideas, research results and insights, as well as fostering dialogue and creating networks between different stakeholders.*
>
> *(Heininen, 2019)*

IASC's vision is in its nature very conducive to cooperation, as shown by the examples of numerous international projects during the period under consideration – that is why there is no detectable axiom of independence in the case of this organization, where each entity remains committed to its goals. Analyzing the effectiveness of its activities, it should be noted that it is dependent on other entities: both state and economic. However, it cannot be said that ISAC is equivalent to state or business entities, and, therefore, it is difficult to talk about ideal symmetry. Scientific research requires large financial outlays, where, despite different goals, States decide to undertake joint research projects, e.g. on researching the bottom of the continental shelf of the Lomonosov Ridge (financed by various States). Therefore, it can be pointed out that the activities of the IASC belong to the axiom of rationality, where the solutions proposed by this organization are subject to negotiations. This is due to the conflict of interests of different groups. On the one hand, we are dealing here with large oil corporations, whose goal is to produce, and which are often focused only on financial profit, and ignore the issues of damage/losses caused by their activities. This reflects the perspective of Nash, where in a short period of time entities deciding on a conflict achieve greater profits from it, but in the long run, the losses are incomparably greater, which will be mentioned in more detail when discussing the activities of NGOs and in the part analyzing rivalry in the Arctic. State entities should make the most of IASC's experience and knowledge. It should be important to find the broadest possible compromise.

Greenpeace and WWF

Apart from IASC, Arctic projects have started to be dealt with by NGOs. Among many organizations, we can mention Greenpeace, which despite many activities in the region has not become an observer in the Arctic Council. It is interesting that Greenpeace actually started its activity in the Arctic. In 1971, a group of founders aboard the boat *Philis Comac* sailed out of Vancouver to protest against the nuclear tests carried out by the US government on the Aleutian Islands, i.e. to the southeast of Alaska. Greenpeace operates in more than 55 countries and is supported by more than 11 million people in the world, of whom about three million regularly finance its campaigns. Therefore, representatives of all the entities of the North American Arctic are very much present there.

Being an NGO, it is fully financed by individual donors but does not accept money from companies or governments. Its main goals are: slowing down climate change, protecting forests, protecting oceans, putting a stop to whaling, opposing the use of nuclear energy and promoting renewable energy, opposing genetic engineering, and eliminating the use of toxic chemicals. Greenpeace offices are largely autonomous in running a joint global campaign tailored to local conditions and seeking the necessary financial support from individual donors (Greenpeace International, n.d.).

Although in the light of the theory of realism or neorealism the role of NGOs is negligible, in the course of more than 40 years of the organization's existence it has managed to achieve many goals. One of Greenpeace's programs was the "Save the Arctic" campaign (Greenpeace International, 2013). On the one hand, this unique region is extremely sensitive and vulnerable to global warming. On the other hand, it is one of the least protected areas of our globe. Although the decreasing amount of ice threatens Arctic fauna and flora, it opens up access to oil and natural gas resources. The possibility of its extraction is an opportunity for financial profit for the oil industry and selected countries. Shell and Gazprom are particularly interested in Arctic resources and seem to care little about the consequences of a possible disaster. It should be noted that governments maintain mankind's dependence on oil at a constantly growing level, instead of introducing regulations that would increase energy efficiency and reduce the share of fossil fuels in the total energy balance of countries, although, admittedly, some progress has been noted in this respect.

Greenpeace mobilizes the community around the world to take action to protect the Arctic, including the establishment of the Arctic Nature Reserve (Greenpeace International, 2014b). The concept of creating an Arctic Nature Sanctuary around the North Pole aims to ban drilling and industrial fishing. The activists are collecting signatures to support this project, which may convince the United Nations (UN) to adopt a special resolution to protect the Arctic. An interesting example of Greenpeace's activity can be the campaign against the Lego concern and Shell. As Greenpeace activists point out, "Shell is polluting the imagination of our children," demanding that Lego breaks off its cooperation with the oil giant. The Internet campaign, conducted through the popular YouTube, was watched by over six million people. In the presented film, the Arctic sinks under a huge amount of oil leakage. Polar bears, people, dogs, and birds drown, and fish perish in the sea. All this is allegedly Shell's doing, which extracts oil in the Arctic and destroys the environment. The film is entitled *LEGO: Everything is NOT awesome* (Greenpeace International, 2014a), and the Arctic in the film is built of Lego bricks and populated with humans and animals produced by the Danish company. In this case, the campaign of the ecological organization has brought concrete results. The

President of the Danish toy company, Jörgen Vig Knudstorp, gave an interview to the magazine *Politiken*, saying that he would not extend the cooperation agreement with Shell. Thanks to the Greenpeace campaign, the contract, which is worth 81 million euro, has expired (Vaughan, 2014). As Annika Jacobson, Greenpeace representative, emphasized, the decision of Lego's boss is a signal for oil companies and an important step toward stopping Shell from extracting oil in the Arctic.

An important event for the protection of the Arctic was the protest, which began on August 24, 2012, involving the occupation of Gazprom's Prirazlomnaya oil rig in the Pechora Sea. It should be emphasized that for Russia, the conquest of the Arctic has been a strategic goal, and the field where Prirazlomnaya is located is estimated to hold some 526 million barrels of oil. The exploitation is of great importance for Russia, which is striving to remain at the forefront of world oil producers. There were six Greenpeace activists on the platform, including the Executive Director of this organization – Kumi Naidoo. They protested against oil production in the Arctic and demanded an immediate stop to oil production in the region (Van Rouveroy van Nieuwaal, 2014). Once again, Greenpeace activists from *Arctic Sunrise* tried to access Gazprom's fuel company platform on September 18, 2013. They wanted to protest there once again against oil extraction in the Arctic. At that time, the Federal Security Service's Coast Guard intervened, as a result of which the Murmansk District Court ordered the arrest for two months of all 30 members of the ship's crew, including 26 foreigners from 18 countries. Initially, the Committee of Inquiry charged them with piracy, for which they were facing 10–15 years in prison in a penal colony. The action was a peaceful protest aimed at stopping the Russian fuel giant from exploiting oil resources in the Arctic. During the voyage, the Russian coast guard fired 11 warning shots at the ship and threatened to shoot at the ship should it not leave. They also demanded the possibility of boarding *Arctic Sunrise* but were refused by the captain. According to Ben Ayliffe, the coordinator of the "Arctic Oil" campaign in Greenpeace International, such a request was unjustified and illegal. The analyzed case from the perspective of balance theory is very interesting. First, Greenpeace chose to create a conflict. In the next round, the Russian Federation also opted for conflict, which from the theoretical perspective should ignite the dispute. However, it should be remembered that they are not the only participants of the Arctic game. The detention of the participants of the *Arctic Sunrise* cruise has activated the international community and gave the incident international publicity. The release of the activists was called for by politicians, artists, NGOs, and over a million people from all over the world, including Nobel Peace Prize winners. As a result, in the face of media pressure, even V. Putin expressed his view in the Russian city of Salekhard during the International Arctic Forum on September 25, 2013, stating that it was absolutely clear that Greenpeace activists were not pirates. The pressure of public opinion and the media coverage of the case have helped to change the charges. On October 23, the committee announced that it had changed the classification of the charges from "piracy" to "hooliganism," punishable by up to 7 years in prison. The Hamburg-based International Tribunal for the Law of the Sea, which ordered Russia to release *Arctic Sunrise*, a ship belonging to the ecological organization Greenpeace, and its entire crew, from Russia, also spoke out against the activists. The verdict was of particular importance, although Russia tried to ignore it. In view of this attitude, on March 17, 2014, Greenpeace sued Russia before the Human Rights Court for the arrest of the crew of *Arctic Sunrise*. The organization demanded compensation and a declaration that the ship's members had been illegally interrogated and detained (Tomala, 2015). In conclusion, Greenpeace's campaign for the release of the Arctic 30 was a success which, as usual, can be perceived from form a variety of way. First, Greenpeace activists who spent more than 2 months in Russian custody after a peaceful protest against oil extraction in the Arctic were released and the Russian side changed its stance, giving in to pressure of the organization and international community. On December 18, 2013, the Duma voted for their inclusion in the planned amnesty. They stressed, however, that "there is no amnesty in the Arctic," and that mechanisms for its protection are needed. Second, Greenpeace succeeded in exposing the public to the dangers of oil extraction in the Arctic. As the activists themselves emphasize, these events may become an important step on the way to building an Arctic Sanctuary and protecting the region in the future.

The presented examples show that Greenpeace is, in light of Nash's theory, an entity that chooses not to compromise or cooperate. Their approach to action is best reflected in the title of the article "We can either all win or we all lose" (Braga, 2020). It is difficult to talk about an axiom of rationality in this case. As indicated in this section, Greenpeace does not back down in its environmental demands and does not compromise. It does not give up its demands either in confrontation with the Arctic States or international companies. It is interesting to note that although its strength on the international arena is quite weak, the means and instruments it uses allow the organization to achieve its own goals. It sets requirements for other participants of the Arctic game as well as shapes ecological standards and norms. Therefore, it is affected by the axiom of independence toward the above-mentioned entities. It does not change its goals and tasks – therefore it is a much-needed entity, balancing the economic interests of other participants in the Arctic game – especially the States that decide to compromise with oil companies. In the light of the theory of balance, it is an entity that will not allow concessions and refuses to accept disregarding ecological, environmental, and social guidelines in the Arctic. The example of Greenpeace also shows that subsequent rounds of conflict do not necessarily translate into a more severe conflict. In these cases, the parties have changed their position (both Russia and Shell) – which is optimistic about the benefits of Arctic cooperation.

Similar in nature to Greenpeace is the WWF, but only Greenpeace, among the NGOs, has observer status in the Arctic Council. WWF includes all Arctic States with the exception of Iceland, which means that, as with Greenpeace, representatives of Canada, the United States, and Denmark belong to this organization. Its mission is to build a future in which humanity will live in harmony with nature. WWF emphasizes that as humanity we are fully dependent on the state of the natural environment, so the organization focuses on saving the most valuable natural sites in over 100 countries on six continents. Another premise is that while thinking globally, we should act locally. Therefore, since 1992, one of the directions of their activities is the Arctic. During the 20 years of the Arctic Program's operation, WWF has built the basis for its Arctic work and directions of its activities in the region. The map of tasks, presented within the framework of this program, presents the scope and diversity of current WWF projects in the Arctic, including: protected areas, conservation, sustainable use, research, fisheries, oil, climate change and resilience, shipping, and governance (WWF, n.d.). The foremost issue is Arctic governance. This is the more global aspect of their work – no matter how good regulations or management systems may be in any one Arctic jurisdiction or sector, that good work can be undone by a neighboring jurisdiction or another sector. WWF has commissioned reports showing where the gaps in governance are and suggesting ways to fix those gaps. They work with Arctic States and Indigenous peoples and advocate for improved Arctic governance, within the Arctic Council and elsewhere. The second area is climate research and communication. As they said: "The big picture is clear: the Arctic is warming, and fast. The smaller picture is harder to understand." WWF is supporting research into processes which are driving and effecting the changes at a regional or community level, both by their knowledgeable staff and by other acknowledged international experts in the field. By communicating this improved knowledge to local and global audiences, they can help local people make decisions about how to manage changing ecosystems. They also supply global decision-makers with information they need to consider in international climate negotiations.

By being an observer in the Arctic Council, WWF has the opportunity to influence the Arctic States to a greater degree. Although WWF notes in its report of 2017 that the Arctic Council has undertaken numerous reviews and reports on the environmental status of the Arctic, it questions whether the recommendations made by the working groups have been implemented effectively in the region. Thus, in 2017, a report assessing the effectiveness of the activities of the Arctic Council and the Arctic States was prepared. It indicates the readiness of the countries in this area to cooperate and the possibilities of achieving the set goals and progress in their actions. Based on the analyses of indicators, WWF indicates recommendations in six areas: conservation, biodiversity, shipping, cooperation on oil spill prevention, preparedness, and response, black carbon adaptation, and ecosystem-based management (EBM) (WWF, 2017). It also assesses the implementation of the recommendations to the Arctic Council and its members in the above-mentioned areas (see Table 3.1).

This example shows the role of WWF in two ways. Firstly, the presented results evidence the scale of the cooperation, including its strengths and weaknesses. Secondly, the results presented in the table show the scale of engagement of the actors in Arctic cooperation and the effectiveness of implementation of recommendations by the Arctic States and the Arctic Council itself. The importance of this document should be emphasized, because in order to speak about the development of environmental cooperation and sustainable development in the Arctic, it is necessary to identify the strengths and weaknesses. Therefore, such an initiative makes it possible to examine the commitment of both the Arctic Council and individual States.

Considering the assumptions of the Nash theory, it can be pointed out that WWF is more committed to cooperation with the Arctic States than Greenpeace. Although its activities are not so spectacular, the examples show the significant impact that this organization has in the international arena. Taking into account the arbitration scheme, in this case we can speak of an axiom of rationality, where the solution of problems will be a part of the negotiation set.

The Arctic Council

Arctic cooperation has not been limited only to scientific research or the activities of NGOs. In 1989, cooperation at the ministerial level was also initiated. Two states came forward with a proposal: Finland and Canada. The Finns considered the institutionalization of this cooperation a priority. They pointed to the need of establishing contacts with the Union of Soviet Socialist Republics (USSR). After consultations with other Arctic States, it was agreed

Table 3.1. Total Score for National Implementation Action Progress in Each Assessment Area.

Arctic States	Conservation Areas	Biodiversity	Shipping	Cooperation on Oil Spills	Black Carbon Adaptation	Ecosystem-based Management
Canada	C	D	C	B	A	C
	6/11	5/17	5/11	19/24	8/8	7/12
Kingdom of Denmark	B	D	C	A	A	D
	7/11	6/17	5/11	22/24	7/8	3/12
United States	B	C	D	A	B	D
	7/11	10/17	3/11	23/24	6/8	3/12
Arctic Council	A	D	A	B	A	D
	1/1	2/6	12/12	7/9	5/5	1/4

Source: (WWF, 2017).

that it would be best and safest to focus on environmental issues, while trying to avoid political issues that could contribute to the project's failure (Berger & True North Strong and Free Inquiry Society, 1989, pp. 54–55). It should be noted that the Finnish concept was intended to expand the circle of countries traditionally considered to be Arctic. It took into account regional interests, as evidenced by the inclusion of the Nordic countries, Sweden, Iceland, and Finland, which were originally not associated with the Arctic. The opposite idea for cooperation was proposed by Canada, which was interested in establishing a broad international cooperation. The Canadians have prioritized the issues of Indigenous peoples and the development of NGOs. To this end, the Arctic Council Panel, consisting of the Canadian Arctic Resources Committee – CARC – and the Canadian Centre for Arms Control and Disarmament, and Innuit Circumpolar Conference – ICC, began work on preparing a project to establish the Arctic Regional Council and prepare a report on the basic principles of cooperation in the Arctic. The Canadians then proposed a draft for the Arctic Treaty (Graham, 1997, p. 53; Pharand, 2019).

These two proposals were quite different. Although the Canadian concept did not gain supporters, the country actively joined the framework of the cooperation proposed by the Finns. This attitude of Canada clearly shows flexibility and willingness to cooperate with other entities. In the light of Nash's theory, it could be assumed that the lack of acceptance for the Canadian proposal would have contributed to the country's withdrawal from cooperation, thus playing only for own benefit in the next stage of the Arctic game. However, this did not happen, and despite everything, the State decided to get involved in the Finnish project. Hence, on June 13–14, 1991, representatives of the Arctic Eight met in Rovaniemi, signing the Declaration on the Protection of the Arctic Environment and the Arctic Environmental Protection Strategy – AEPS (Arctic Environmental Protection Strategy, Declaration on the Protection of the Arctic Environment, 1991). The Strategy set out the directions of development of legal regulations for the protection of the Arctic environment. These countries, due to their location around the North Pole, recognized their special role in strengthening the mechanisms for environmental protection. Emphasis was placed on monitoring of the state of the environment and identification and harmonization of the internal and international measures for its protection. The work of AEPS focuses on projects aimed at protecting the Arctic environment, including: protection of Arctic ecosystems, Indigenous peoples, striving to maintain the high quality of the environment in the Arctic and sustainable use of its natural resources by local inhabitants, using the traditional knowledge of the Indigenous peoples of the North in environmental activities, regular review of the state of the environment in the Arctic region, and the recognition, reduction, and consequent complete elimination of pollution of the Arctic environment (Stokke & Hønneland, 2007, pp. 1–3; Tomala, 2016, pp. 130–131).

It should be noted that the Strategy highlights the most important Arctic problems, i.e. persistent organic pollutants, oil pollution, heavy metals issues, noise, radioactivity, and environmental acidification (Huebert, 1998, pp. 37–38). The main mechanisms corresponding to each type of pollution have been identified, and actions to be taken to improve the situation in the region determined.

Four working groups were established to carry out these tasks (Murray, 2014, pp. 27–28):

(1) The Arctic Monitoring and Assessment Program (AMAP) monitors and assesses pollutants in the Arctic. They produce assessment reports and recommend responses.

(2) The Conservation of Arctic Flora and Fauna (CAFF) exchanges information on how Indigenous peoples use and interact with Arctic species.

(3) The Emergency Prevention, Preparedness, and Response Working Group (EPPR) provides a framework for cooperation in response to environmental emergencies.

(4) The Protection of the Arctic Marine Environment (PAME) group conducts activities related to the protection and sustainable use of the Arctic marine environment.

The first time representatives of the Arctic States met in Copenhagen in 1992. Since then, annual meetings have become a practice, reviewing the progress of the AEPS Action Plan. It appears that it is indeed the right direction as AEPS's activities improve the state of the environment in the Arctic. Its monitoring has made the international community and Arctic States aware of the state of environmental pollution, and the risks associated with warming and melting of the Arctic ice cap.

Moreover, the fact that the States decided to control the Arctic environment enabled them to take corrective actions in the area of fauna, flora, and marine environment. The setting of cooperation priorities and identifying problem areas should be assessed as achievements and benefits of cooperation at this stage in the light of J. Nash's theory. However, it should be noted that at that time it was difficult to point out the effects achieved by the States in relation to the undertaken actions. It should be emphasized that it is not easy to show the environmental threats in the Arctic and the results of actions because of the need to create the right tools to study the state of the ecosystem in the Arctic and to monitor nature in the long term. In the past, monitoring practices in the Arctic were fragmented and incomplete. As indicated above, it took more than several decades for such tools to be developed. In the future, they will allow monitoring and comparing the progress of cooperation within the Arctic Council. The 2004 Arctic Climate Impact Assessment (ACIA, 2004) recommended that long-term Arctic biodiversity monitoring be expanded and enhanced. In response, two of the Council's working groups – CAFF and AMAP – examined the report's findings and developed follow-up programs that address key projections for the future of the Arctic.

An important event strengthening the Arctic States' cooperation for the environment was the ministerial meeting held one year after the Earth Summit, i.e. in Greenland's Nuuk in 1993. The issues discussed were a direct result of the provisions of the United Nations Conference on Environment and Development (United Nations. General Assembly, 1992, p. 133). In Nuuk, it was decided to further develop the structures of Arctic cooperation and its extension to the area of sustainable development (The Nuuk Declaration on Environment and Development in the Arctic: Report, 1993). At the same time, it was decided to establish a Task Force on Sustainable Development and Utilization (TFSDU). On the initiative of Denmark and Greenland, it was also resolved to establish the Indigenous People's Secretariat – IPS.

Within the framework of the AEPS, two more ministerial meetings were held, the first one in the Canadian city of Inuvik in March 1996 and a year later in June in Alta, Norway, where it was adopted to complete the Rovaniemi process and to include the AEPS with its working groups and the IPS in the activities of the newly established organization – the Arctic Council.

The Arctic Council was founded on September 19, 1996, with the status of an intergovernmental organization. It is strictly exclusive, which means that countries not in geographical contact with the Arctic do not have the right to participate in its deliberations (Arctic Council, 1996). There are three categories of participants. First of all, these are the Arctic countries, the only ones with the right to make decisions and to have the status of member states. Technically, there is no voting in the Arctic Council; the body makes decisions by consensus. Here we can distinguish the Arctic Five States: the United States, Canada, Denmark, Norway, the Russian Federation, and the Arctic Eight, consisting of the Arctic Five and Finland, Iceland, and Sweden. This means that the three States of the North American Arctic can play a key role in deciding on issues in the region. Their coalition can play a role in building an advantage over the Russian Federation. Other participants are Indigenous Peoples' Organizations – IPOs, described as Permanent Participants. The last group of the Arctic Council are observers. This group includes accredited non-Arctic States and global and regional NGOs (The Arctic Council, 2020).

As emphasized during the first ministerial meeting of the members of the Arctic Council, its main objective is to provide the means to promote cooperation, coordination, and interaction between the Arctic States, involving Indigenous peoples. Its area of interest is environmental protection and sustainable development, such as conducting comprehensive scientific research, monitoring the consequences of climate change and helping Indigenous peoples, opening new shipping lanes, and managing natural resources.

The WWF Arctic Council Scorecard is an assessment tool of the implementation of Arctic Council commitments on biodiversity, conservation, EBM, black carbon emission, oil spills prevention, and shipping. Based on the rating system, it addresses the implementation of the action taken by eight Arctic States individually, and the Arctic Council, in categories from A to D, where A represents the best performance and D the worst. The first Scoreboard of 2017 indicated that the Arctic Council in conservation, shipping, and black carbon delivered its commitments, but little progress was observed in biodiversity and EBM. Canada, Denmark, and the United States successfully established an administrative system for oil spill response and addressed black carbon emission, with early action taken to reduce emission, but they generally failed in safeguarding biodiversity, reducing air emissions from Arctic marine shipping, and implementing EBM. Although, the states identified the areas for conservation, national implementation progress was limited. (Arctic Council Scorecard, 2017). The 2019 Arctic Council Conservation Scorecard indicated that Arctic States did not fulfill their role as the prime steward of the region, although they improved using EBM in their planning. They especially made slow progress in safeguarding biodiversity and protecting conservation areas and did not show willingness to recognize Indigenous peoples as full partners in managing the region. The North American Arctic States received the worst scores in the regulation of Arctic marine shipping (Arctic Council Scorecard, 2019).

It is worth noting that the Arctic States' cooperation has not been limited in benefits to those countries alone. The importance of protecting the Arctic environment attracted the attention of many non-Arctic States and NGOs. The improvement of the environment has a global dimension and concerns all countries in the world. It cannot be overlooked that the strengthening of environmental cooperation is not necessarily for purely altruistic reasons. The cooperation between the hitherto competing entities has contributed to the development of cooperation in other areas, such as international trade, development of diplomatic relations, and others.

In assessing the role of the Arctic Council in the light of the theory of equilibrium, it should be noted that the Council is rather reserved in terms of developing and expanding cooperation. This is evidenced, inter alia, by the roles of the participants in the Arctic Council. Firstly, it is proven by the fact that such organizations as the European Union (EU), Greenpeace, International Hydrographic Organization (IHO), and Association of Polar Early Career Scientists still strive to gain the observer status in the Arctic Council, although they, with their knowledge and experience, could undoubtedly enrich the activities already being conducted. Cooperation with the EU would be particularly beneficial for the Arctic region. It should be mentioned that cooperation of entities in the broadest possible scope will allow achieving the objectives of sustainable development, since the Arctic is polluted not only by entities concentrated around the North Pole. The EU experience in meeting the commitments of climate policy and the concept of sustainable development could contribute to improving the situation in the Arctic. Another issue is Indigenous peoples, who have not gained more power of decision-making within the Arctic Council during the period of its operation. Thus, we can see a lack of trust in cooperating partners, which manifests itself in a rather conservative approach, perhaps playing it safe in terms of the problems to be solved. This is manifested by the choice of the Finnish concept, more general and irrelevant to any specific area of cooperation, and the nature and status of the Arctic Council, which has not been given a mandate to manage the Arctic on its own. The lack of a broad range of non-Arctic States that do not have the right to decide within the Arctic Council is also a signal that the Arctic players – especially the Arctic Five – have profound interests in the region. The above-mentioned premises allow to state that for the Arctic Council the axiom of symmetry is typical, in which negotiation problems determine the outcome that does not distinguish between the parties concluding an agreement. This axiom reflects the equal potential and negotiation skills of the players of the Arctic Five, but it is a threat that has often blocked the development of this cooperation for other entities.

Indigenous Peoples Organizations and Circumpolar Council

The people of the polar regions see themselves as neighbors of nonpolar regions in the world. Their concerns are raised by the impact of global problems on their own living environment. This applies primarily to whaling, hunting, or mining the natural resources, which have been the foundation of their lives for hundreds of years. The people of the Arctic have adapted to life in this region, despite unfavorable climate conditions. Through close contact with nature, they have created their own local culture and rich tradition. Drawing on natural resources, they managed to maintain a strong hunting and gathering tradition.

However, due to the increasing influence of external factors on the life of the local communities, such as colonization, Westernization, Europeanization, or globalization, the life of the Indigenous people is changing (Myrdal, 1957; Spandau, 1968, p. 282). These processes are often the result of the political play of those who, by portraying the Indigenous peoples through the prism of their hunting prowess, "whitewash themselves" in order to justify their exploitative, environmentally destructive actions, even though the Natives' traditional knowledge is essential for proper governance in the Arctic (Arruda & Krutkowski, 2017, p. 515).

These processes contribute to the abandonment of traditional occupations and the way of life of the North's Indigenous peoples, which contradicts the sovereign status they should have (Arruda & Krutkowski, 2017, pp. 278–279). In turn, this raises the issue of the imbalance of power and importance of international players representing nation-states, oil companies, and the Inuit. Awareness of these relationships contributed to the establishment of an NGO that would act on behalf of the Natives in the international arena, defending their interests and the right to preserve their own culture and traditions. Its aim was also to protect economic interests connected with the growing interest of international subjects in the possibilities of using these lands for the development of the mining industry. The ICC, which represents the peoples who live in the polar circular areas, has undoubtedly become such an institution. Overall, there are six organizations in the Arctic that represent Indigenous peoples, four of these organizations represent the North American Arctic: Arctic Athabaskan Council (ATC), Aleut International Association (AIA), Gwich'in Council International (GCI), and ICC. The first one, established to defend the rights and the interests of 42,000 American and Canadian Athabaskan people, promotes the balancing of environmental protection with economic sustainability. The second one represents the Indigenous peoples of Aleut descent in the United States and the Russian Federation. AIA was formed to address environmental and cultural concerns of the extended Aleut family. The GCI represents 9,000 Gwich'in in Alaska, United States, and the Yukon and Northwest Territories in Canada. Its mission is to amplify the voice of the Gwich'in Nation on issues of sustainable development and the environment in international fora.

The fourth organization, the ICC, deserves special attention as a major international organization representing 180,000 Indigenous people. The creation of the ICC resulted from the need to speak with one voice on Arctic issues in the international arena. The most important objectives of the ICC (ICC, n.d.) are to strengthen the unity of the Inuit people living in the polar regions, to promote the rights and interests of the Inuit people at the international level, to develop and encourage the international community to develop a strategy that takes into account the need to protect the Arctic environment, and to seek a full and active partnership in the political, economic, and social development of the polar regions.

Greenland's penultimate presidency of the ICC is noteworthy. From June 28th to July 2nd, 2010, the 11th General Assembly of the Inuit from Alaska, Canada, Greenland, and Russia took place in Nuuk, Greenland, to celebrate the 30th anniversary of the ICC under the slogan *Inoqatigiinneq* – sharing life. This meeting was intended to follow up on the first Inuit Conference (ICC) in Alaska in 1977, where Eben Hopson Senior invited Inuit people from all over the Arctic to celebrate the unity of the Inuit and commit everyone to international joint action. The participants of the meeting attempted to summarize the 30 years of the Council's activity, during which the challenge was to build cooperation on the international scene, based on the professional advice of the Inuit. As a result, this contributed to the understanding of the unique status and reputation of the ICC within the Inuit Organization (IPO) both in member states and abroad. The main slogans of the Assembly were (ICC, 2010):

> *Remembering that the respectful sharing of resources, culture and life itself with others is a fundamental principle of being Inuit, and is the fabric that holds together as one people across four countries.*

The delegates pointed out that there is a need to record the changes taking place in the Arctic, including the measurement of ice melting in this area where there are serious environmental risks. Their concern was that the history of the Inuits has been a constant challenge to adapt to change in order to find livelihoods and that influential powers, agencies, and multinationals are increasingly interested in exploiting the Arctic's marine environment and its resources. These arguments, which were cited during the Council's session, contributed to the conclusion that the Inuit, as Indigenous inhabitants of vast areas of the Arctic, including its coastline, has rights and responsibilities for the management and future of the marine environment toward present and future generations and toward the whole humanity.

In turn, the theme of the Canadian Presidency of the Council was the message "One Arctic, One Future." The General Assembly held in Inuvik (Canada) on July 21–24, 2014, adopted the Kitigaaryuit Declaration

(Kitigaaryuit Declaration, 2014). Issues discussed included the development of ICC cooperation in international fora, safety of maritime transport and fisheries, food safety, education, language, and tradition.

As President of the ICC in Canada, Duane Smith stressed, the goal of Indigenous peoples is to actively participate in the natural resources policy of the Arctic States. Terry Audla – Inuit leader – pointed out that the Inuit people should become employers, not just employees in Arctic companies, because they own these areas and strive for economic development of their areas.

The latest chairmanship of the ICC was held by the United States. The meeting was held in Utqiaġvik, Alaska, on 16–19 July, 2018, with the theme of "Inuit – The Arctic We Want." In the conclusions, the delegates emphasized the interconnectedness of the indivisible rights of the Inuit, which were confirmed in the UNDRIP – the United Nations Declaration on the Rights of Indigenous Peoples – as an international human rights instrument, as well as in other legal and political instruments and mechanisms, including legislation and agreements on land rights and on self-government, and intergovernmental and constitutional arrangements. The most important goals include Food Security, Families and Youth, Health and Wellness, Education and Language, Indigenous Knowledge, Sustainable Wildlife Management, Environment, Sustainable Development, Communication, and Capacity Building (Utqiaġvik Declaration 2018, 2018).

When evaluating the ICC's activities through the prism of Nash's theory, it should be noted that it is a collaborative organization, but at the same time, this particular player is gaining more and more strength in the Arctic. The change that has taken place in this respect over several decades is evidenced in the Address to all Indigenous peoples of the Arctic, governments of Arctic nations, and international organizations at Usinsk, Russia, on August 14–16, 2012:

> *We the Peoples of the North have for too long experienced the oppression of our Peoples and the barbaric destruction of our land. It is time that we join forces and demands that the oil companies and the Arctic States change their path and start to listen to the voices of the Indigenous Peoples residing in these lands. The Peoples of the North will no longer be bought with dimes and cents to stand silently by while the oil companies destroy our native land.*

The ICC sets goals, and it has shown it can achieve them. It should be emphasized that the core of activities is focused on the cooperation of the North American Arctic countries and Russia. The organization takes into account in the negotiation area both the priorities of ecological and business

NGOs. Its activities are focused on finding the most effective solutions, as evidenced by the leading themes implemented by successive presidents of the ICC. The essence here is the concept of sustainable development, which takes into account tradition, but also sets modern priorities for the people of the region, and cooperation within the UN. Thanks to it, the organization as well as the Indigenous peoples gain the rights to manage the Arctic, which enables them to slowly become independent from state entities or international enterprises.

ROUND TWO: RIVALRY AND CONFLICT

The Background

The first round of the North American Arctic games confirms the change in relations that took place between the subjects in the early 1990s. It seemed that the benefits of cooperation and the development of governmental organizations and NGOs that are involved in Arctic cooperation would consolidate the new hand. It should be remembered, however, that the dispute over North American Arctic territories did not erupt in the twenty-first century, but its roots go back to 1909, when Canada filed claims to the waters of the Arctic Ocean. Despite this, the status of the Arctic had not been regulated by any international agreement, as the Antarctic, but the number of entities interested in the rights to these territories has increased.[1] It is, therefore, difficult to analyze disputes and conflicts between the North American Arctic States without also focusing on other participants in the Arctic games. This section focuses on those conflicts and rivalries in which the North American Arctic States participate. The question can be asked whether they are important enough to squander the cooperation achievements mentioned in the previous section.

The dispute over the Arctic territories has a strategic and economic background. The struggle for raw materials takes place between the United States, Canada, Russia, Norway, and Denmark, i.e. the so-called Arctic Five. Together with the three remaining members of the Arctic Council, the territorial claims of these States reach nearly 90% of the territory of the High North. This is because more and more often the Arctic States submit their

1 Status of the Antarctic is regulated by the system of international agreements, collectively known as the Antarctic Treaty System (ATS). The Arctic and its status differ considerably as explained in the Chapter one.

claims to this part of the world, the regional cooperation established in the 1990s recedes into the background, and the Arctic Council begins to transform from an organization of a consultative character into a political body, shaping strategic directions of development in the High North.

The following events contributed to the departure from cooperation of the Arctic States. Firstly, in 2008, the Geological Survey of the United States developed a report on the statistical probability of occurrence of natural resources in the Arctic, including the key oil and gas resources for economic development (Gautier et al., 2009, pp. 1175–1176). Optimistic conclusions about the attractiveness of this region have increased the interest of international relations subjects in potential profits from the extraction of energy resources and the development of maritime transport.

The second issue was the understanding of climate change processes. The warming climate contributes to the disappearance of the ice cap in the Arctic Ocean, and, thus, the process facilitates the exploitation of deposits located in the region. It is worth noting that it is the results of scientific research, presented during the Intergovernmental Panel on Climate Change (IPCC) in 2001 and the 2004 Impacts of Warming Arctic: ACIA Overview Report of the Arctic Council that influenced the behavior of different categories of actors (both regional and nonregional) on the policy toward the Arctic. For the first time in history, the third IPCC report showed the special situation in the Arctic and its role in the global context (IPCC, 2001). These conclusions were also confirmed by the 2004 report of the Arctic Council, which showed a wide range of changes taking place in the region related to climate change (ACIA, 2004). The latter report presents the most comprehensive scientific analysis on climate change, leaving no doubt that we as humanity are witnessing a profound transformation. The IPCC calls for action and changes in countries' climate policies, indicating the transition to a low-carbon economy (IPCC, 2014c).

The third reason for the emergence of conflicts in the Arctic region is the lack of international regulations relating directly to the Arctic. Currently, the only international document is the United Nations Convention on the Law of the Sea signed in Montego Bay on December 10, 1982. However, it does not directly specify the division of the Arctic territories, which allows for much diverse interpretations of the regulations, and this, in turn, brings the conflicting parties (the States) to trying to prove their rights to the disputed territories in the most beneficial way for them.

The above-mentioned premises instigate the States to emphasize their rights to the Arctic territories. The actions taken nowadays are rather symbolic or demonstrative in nature, but they point to the conflict rather than to cooperation between the countries in question. Some of these disputes have already been resolved, but there are still examples of ongoing or potential conflicts. The following section, therefore, focuses on those examples that concern the North American Arctic States.

Danish–Canadian Dispute over Hans Island

Hans Island is located in the Kennedy Channel in the Nares Strait. Its area is only 1.3 km^2, and the terrain is practically barren. Thus, geographical conditions are not sufficient for it to be inhabited. Although the dispute about the isle is not important from the economic perspective, it has a symbolic meaning. The island is located between Denmark's Greenland and Ellesmere, the latter being under Canadian jurisdiction. In the 1970s, 1980s, and 1990s, one could observe mutual attempts to convince the other about the rightful ownership. Canada was the first one to make claims in 1971, when Greenland negotiated the sea border in the Nares Strait with Denmark. The negotiations were concluded on December 17, 1973, with the signing of the delimitation agreement of the Continental Shelf, which entered into force on March 13, 1974 (Stevenson, 2007, p. 266). In the light of the agreements, the Nares Strait was divided in accordance with 127 landmarks. However, the agreement did not take up the issue of Hans Island, which in effect meant that the border line crossed the isle almost exactly in the middle (Agreement between the Government of the Kingdom of Denmark and the Government of Canada relating to the delimitation of the Continental Shelf between Greenland and Canada, 1973). In order to gain rights to the Island, Canada in the 1980s took further actions. It gave permission to Dome Petroleum, a fuel industry company, to conduct scientific research. The purpose of this research was to examine the impact of melting icebergs on the island and, in fact, to study the impact of moving ice masses on oil platforms. The Danish Government had not been informed about this by the Canadian side. It was only the Greenland historian Kenn Harper who informed the authorities when he came across a Canadian researcher representing "Hans Island N.W.T." during the expedition (the Northwest Territories of Canada). This event was protested against by the Danes, and then it was they who claimed their rights to Hans Island. In 1984, the Danish Minister responsible for Greenland, Tony Høyem, landed on the island in a helicopter, marking the island with the Danish flag and leaving a message "Welcome to Denmark." In addition, the minister buried a bottle of Danish brandy on the island – which opened a new chapter in a diplomatic dispute called the "Whisky War." In 2005, Canada wanted to demonstrate its presence. At that time, the Canadian Minister of Defense, Bill Graham, placed the flag on the island, and in addition, a Canadian military patrol had installed an Inukshuk stone sculpture there a week earlier. As Peter Taksoe-Jensen said when he was legal adviser to the Danish Foreign Minister,

> *When Danish military go there, they leave a bottle of schnapps. And when [Canadian] military forces come there, they leave a bottle of Canadian Club [whisky] and a sign saying, "Welcome to Canada".*
> *(Byers, 2013, p. 13)*

The ceremony of leaving a bottle of alcohol was repeated many times by both sides. The attitude of the Canadians met with not only Danish but also Greenlandic criticism stating that the island is occupied by Canadians. In addition, it was emphasized that the Danish–Canadian dispute should be resolved by independent experts (Guo, 2006, p. 135). On September 19, 2005, an agreement was reached at the UN session in New York. The Ministers of Foreign Affairs of Denmark and Canada declared that the dispute over Hans Island would be resolved. The parties also agreed to keep each other informed of activities on the island. The agreement can be described as an improvement rather than a resolution of the conflict (Canada and Denmark Issue Statement on Hans Island, 2005). There is still no solution to which country the isle belongs to, and neither party has yielded its claims, bearing in mind also other disputes that are taking place in the Arctic about much more important and energy-rich areas of the region.

The dispute over Hans Island is very interesting from the perspective of Nash's theory. In fact, it can be described as a demonstration of the strength of one party, which was not left without an answer from the other party. It is a kind of testing the tolerance limit of those involved in the conflict. This is confirmed by the research of successive rounds of the game, in which no player wanted to give up his/her/its/their conflict strategy. It is also worth emphasizing that this is all the more intriguing because scientific studies do not indicate the benefits of winning, while the losses resulting from giving up cooperation in environmental or scientific dimensions are measurable. Moreover, this conflict is described as the most friendly border dispute in the world, often described humorously in the daily press (Bender, 2016). Indeed, the importance of multilateral cooperation, which has not been weakened by Canada's bilateral conflict with Denmark, should be stressed. However, it is difficult to state that in the future whether the subjects which have more that unites rather than divides them will decide to strengthen bilateral cooperation and friendly relations on Hans Island, all the more so as Canada wants to show its strength in the international arena, bearing in mind also other disputes it has with both the United States and other Arctic players. It is also worth emphasizing the importance of Greenland in this dispute, whose voice contributed to weakening the conflict, but it is worthy of note that the island is a weak international player.

Canada's Dispute with the United States over a Portion of the Beaufort Sea

The subject of the dispute between Canada and the United States is the area located in the Beaufort Sea, much larger than Hans Island, because it measures 6,250 square nautical miles. The conflict is based on the fact that Canada believes that the boundary should be along the 141 degree meridian length – according to the sector theory – while the United States maintains that the boundary should be set at equal distances by setting perpendicular lines of up to 200 nautical miles, which potentially make up the Exclusive Economic Zone. The history of the area is important in this case. The disputed situation dates back to the time when Russia and the United Kingdom – the signatories – contracted a border treaty in 1825. The legal status of the area changed in the following years due to two events. First, in 1867, Alaska was bought from Russia by the United States, and second, in 1880, Canada took over the rights from Great Britain. Canada recognized the border as a line running along the 141 meridian, in its prolongation as far as the frozen ocean (Baker & Byers, 2012, pp. 71–72). The argument in favor of such an interpretation of the law was the border between Yukon and Alaska, from which the border between the United States and the USSR was drawn.

Although in the late 1970s these countries intended to reach a compromise, the matter is still unresolved. The reasons for this are to be seen in connection with forecasts of oil and gas production potential in the region. Byers (Byers, 2014), Gautier (Gautier et al., 2009), and Bird (Bird et al., 2008) in their works recall the US Geological Survey calculations that 83 billion barrels of oil and 44 trillion cubic meters of natural gas are located north of the Arctic Circle. This amounts to approximately 30% of the world's undiscovered gas and 13% of the world's undiscovered oil, which would meet the global demand for oil for three years and gas even for 14 years. As David H. Gray points out,

> *The offshore hydrocarbon potential is definitely identified and much exploration work has been conducted in the area near Tuktoyaktuk on the Canadian side. Oil is being produced commercially from*

> *wells on land near Prudhoe Bay, Alaska and sent south through the trans-Alaska pipeline.*
>
> *(Gray, 1997, p. 63)*

The Prudhoe Bay area is home to the largest known American oil field, located in the west of the disputed area, and production from this field accounts for approximately 25% of total US oil production. Canada has an interest in the Amauligak field, and extremely rich deposits are also estimated in its vicinity (Fogarassy & Wilson, 2007).

In the 1960s, Canada issued permits for oil and gas exploration in the disputed area and then included this territory in a 1970 document, the Arctic Waters Pollution Prevention Act, which aimed to prevent pollution of the Arctic waters adjacent to the mainland and islands of the Canadian Arctic (Department of Justice Canada, 1985). Although these events did not provoke any response from the United States, when the Canadian side began issuing more drilling permits, and on January 1, 1977, declared a 200-nautical mile-wide "Exclusive Fishing Zone 5" based on the 141 meridian in the disputed areas (Government of Canada, 1977), it was only on March 1, 1977, that the American side protested against Canada, proclaiming the 200-mile Fishery Conservation Zone. For that, the United States used an equidistance line based on the low water line (Gray, 1997, p. 62). In the late 1970s, both sides tried to resolve this dispute. It was also related to other conflicts, including the need to agree on the Bay of Maine issue. The Canadian side was willing to agree on a compromise to suffer some territorial losses in the Beaufort Sea, with gains elsewhere. As Gnaś points out, it can be assumed that this was the Gulf of Maine (Gnaś, 2015, p. 42). However, there was an impasse during the negotiations because each side feared losses in the case of other delimitations. As a result, the case has been referred to the International Court of Justice and is still unresolved, although, as Baker and Byers point out, the matter is well managed (Baker & Byers, 2012, p. 73).

The conflict does not exclude cooperation in this case. As Lawrence Cannon points out, in a press release describing why the two countries were cooperating on the joint mission, "Canada and the United States need this data, both to delineate the continental shelf and to assist in the eventual resolution of the Beaufort Sea maritime boundary dispute." The research was aimed at studying the surface of the seabed. Although they should focus on the boundary areas, as Jacob Verhoef explains, they focused on the Canadian and American economic zones. It can be concluded, therefore, that neither party is certain of the rights to the disputed area, so that far-reaching caution can be found in the actions of corporations and scientists. This is evidenced by the attitude of the investors, who mostly resigned from participating in tenders offered by both the Canadian and the American side, respectively.

Analyzing the dispute over part of the Beaufort Sea in the light of the theory of equilibrium, it should be noted that the parties of the conflict maintain a passive attitude. Neither wants to compromise, and they appear to come closer to conflict than to cooperation. However, this does not change the fact that such cooperation between the parties does happen. Therefore, in this case it is difficult to confirm Nash's theory. Also, the issue of successive rounds of negotiations has not been observed in this case, as sequential rounds tend to intensify the conflict. The most probable scenario seems to be a resolution of the dispute on the supranational level, rather than an intensification of the conflict between the States.

The Northwest Passage

The dispute over resources in the Beaufort Sea is not the only one in which the United States and Canada are involved. The Northwest Passage is just as complicated. It is a waterway located between the Davis Strait at the Baffin Sea at the entrance to the Lancaster Sound from the East and the Bering Strait from the West. It is an Arctic seaway from Europe to East Asia, leading from the Atlantic Ocean to the Pacific, and is about 900 nautical miles long. Seven variants of this route are possible, including two main ones. The most important players in this game are the United States and Canada, but it should be remembered that they are not the only ones interested in the freedom of navigation in this area (Gabryś, 2016, pp. 158–160).

The increase in interest in the Northwest Passage is due to climate change. Glacier melting favors both the mining industry (see above) and the development of Arctic shipping. However, this is despite the worrying situation concerning the Arctic environment itself. As Walsh et al. states, "The consolidated database shows that there is no precedent as far back as 1850 for the 21st century's minimum ice extent of sea ice on the pan-Arctic scale" (Walsh, Fetterer, Stewart, & Chapman, 2017). Secondly, this road allows shortening the route through the Panama Canal significantly. It also gives more transport possibilities, so in the history of shipping one can find repeated attempts to take the challenge of crossing the route between the Pacific and the Atlantic Oceans.

Although negotiating the Northwest Passage is conditioned on the reduction of the ice cap to the extent that it can be used by ships that do not have adequate reinforcements in the structure to overcome the ice, this is not the only obstacle limiting taking advantage of this route. The problem is the legal and political status of this passage and the lack of final international regulations on this issue (Roach, Smith, & Roach, 2012, pp. 23–24).

Canada indicates that these waters have been given the status of internal waters, thus denying the rights of other countries to sail freely in the area. When the above-mentioned supertanker *Manhattan* accompanied by a Canadian icebreaker crossed the northwestern route, the United States recognized this passage as international waters. This, in turn, triggered a protest on the part of the Canadian side, and the then Prime Minister of Canada, Elliot Trudeau, announced that the waters in the Arctic Archipelago are the country's internal waters over which Canada has sovereign power. Analyzing further negotiations between the two countries, it is clear that a similar strategy was adopted by Denmark and Canada in the dispute over Hans Island. Here we are dealing with a similar ceremonial, where each side demonstrates its position and its opinion on the status of the Northwest Passage without trying to find a consensus. Canada is using the 1970 Arctic Waters Pollution Prevention Act, which is being challenged by the United States, pointing to Canada's illegal interference in international law and freedom of navigation. Again in 1973, Canada announced that the Northwest Passage was its internal route, which the United States opposed. These bilateral skirmishes, following Suzanne Lalonde and Frederic Lasserre, can be described in a nutshell as a situation in which "For decades, Canada and the United States have been agreeing to disagree on the question" (Lalonde & Lasserre, 2013, p. 28). As a result of these diplomatic games, in 1985 there was a kind of trial of strength or a tug of war between the subjects. At that time, the United States informed Canada of a plan to take the Polar Sea icebreaker through the Northwest Passage and invited Canadians to join the venture. The Canadian side responded diplomatically, appreciating the Americans' gesture, but at the same time reminding that the crossing is its internal waters. The United States reported that it did not share this position, which Canada acknowledged with deep regret. Eventually, despite the "diplomatic chess," a joint project involving the heavy Canadian icebreaker *John A. Macdonald* (Roach et al., 2012, pp. 320–324) was carried out. It is important to note that the United States is not the only country that disagrees with the policy that Canada is forcing, which leaves Canada alone in its claims, and, as a result, has contributed to the development of an agreement between the parties of the conflict. In 1988, Brian Mulroney, on behalf of the Canadian government, and Ronald Reagan, on behalf of the United States, signed the Arctic Cooperation Agreement under which the US Coast Guard may use this route after notifying Canada, but Canada cannot refuse such consent. Although such arrangements have been made, the United States has often deliberately failed to inform Canada of planned movements, which still means there is no agreement on the disputed areas.

This example of a game between the United States and Canada shows the strategy adopted by the players in the subsequent rounds of the Arctic Northwest Passage. It is difficult to confirm here the theory that the countries exacerbated the situation in the subsequent rounds, but it is also impossible to indicate that they were able to reach an agreement.

In fact, all three analyzed cases show a different kind of game than the one proposed by Nash, in which one can talk about balancing the interests of the subjects, diverse forays, testing the strength, and checking the opponent. This is due to the cost–benefit analysis of the conflict or cooperation strategy. Therefore, it can be pointed out that the examined countries will continue this type of game until an international settlement is reached in each of the discussed cases. In this context, international cooperation and the involvement of the North American Arctic actors in the most important problems of the region, i.e. environmental protection and sustainable development, have become of ultimate importance.

ROUND THREE: BETWEEN COOPERATION AND COMPETITION

The Background

Although international organizations can and should play a key role in the governance of the Arctic region, this is unlikely to happen, as evidenced by the status of the EU or environmental organizations such as Greenpeace in the Arctic Council. Building coalitions of cooperating States will be of key importance in this context, which will enable the balance between mining policy and environmental protection. Here the importance of a coalition, which could have been created, but never came into being, by the North American Arctic States, becomes apparent. The importance of the two-track cooperation will be particularly important in this case – Canada with Denmark and Denmark with the United States, which may be of key importance in order to maintain a counterbalance to cooperation with, e.g. Russia, Norway, or other non-Arctic countries, such as China. The question arises: why can these coalitions be of key importance in the system of international relations? It can be hypothesized that the indicated coalition of the North American Arctic States would balance the interests of other Arctic players, especially Russia's actions in the international arena.

In order to understand that the Arctic is not only a subject but also an object of the Arctic games, it is necessary to realize its importance in the international arena. It is not only a reservoir of natural resources, but also has other, strategic importance for State entities. From the geopolitical point of view, the Arctic is the only reliable and certain way for Russia into the ocean, and thus into all the countries of the world, and that is through the Northern Sea Route. The Arctic is also of military importance, since only through the Arctic ionosphere can communication be influenced, and its disruption is the first task during war. It is, therefore, a strategic area, control over which determines both economic and national security issues. In this context, the two-block conflict between the Western block – symbolically represented by the United States – and the Eastern block – represented by Russia – is renewed.

Russia is a key player in the dispute over Arctic territories. The Russians are preparing forces to strengthen the protection of the northern borders, infrastructure, railroads, and roads. They are expanding the navy, entrusting it with the task of protecting the economic interests of Russia and Arctic resources (Burgess, 2010). In the High North, they create Arctic subdivisions, develop military infrastructure, and strengthen borders. The Russian Federation is also intensifying its activities in the Arctic Council and in the entire region. As press reports indicate, it is preparing to occupy the Arctic, taking into account both peace and war scenarios in its plans. However, no radical political solutions should be expected. Russians seem to build their policy slowly, but in such a way that the final territorial solutions will come when they are technologically ready (Baev, 2008, 2018). This scenario is confirmed by Åtland, who noted an increase in aggressive Russian rhetoric already in 2010 (Åtland, 2017), and the culmination of the aggression in August 2018 is noted by Mathieu Boulègue when Defense Minister, Sergey Shoygu, announced that "competition in the Arctic could lead to potential conflict" (Boulègue, 2019, p. 6).

Vladimir Putin's return to the post of president contributed to the militarization of the Russian part of the polar circle. The Russian plans for the Arctic are evidenced by his declaration made during the third International Forum "Arctic – Territory of Dialogue," held in Salekhard on September 29, 2013. The theme of the forum was ecology and health of the inhabitants of the Arctic region. In Salekhard, 375 experts and representatives of relevant ministries of Russia and other Arctic States gathered. Among the delegates of the Forum were the Presidents of Finland and Iceland, as well as the Russian leader Vladimir Putin. In his speech, he stated that a new chapter in history was opening up today for the resource-rich Arctic. The Arctic regions of Russia are actively researching and exploring new deposits of gas, oil, and other mineral resources, new transport and energy facilities are being built, and the Northern Sea Route is being revived. Working in the harsh conditions of the Arctic is extremely difficult, requiring both serious financial contributions and truly unique technological solutions. It is obvious to us that the priority, the key principle of exploiting the Arctic, should be nature conservation, ensuring a balance between economic activity, human presence, and the state of the environment. According to Putin, the Arctic has a chance to "become a major source of energy extraction in the future." He predicted that active exploitation of the Arctic would begin no sooner than in 5–7 years' time, and maybe even 10 years' time. At the same time, extraction activities in the Arctic are associated with an increased environmental risk. During his speech at the Forum, Putin stressed that only companies using the most advanced technologies would be allowed to work in this region. This is important because an oil spill in this region would become a much more serious problem than, for example, an accident on the Deepwater Horizon platform in the Gulf of Mexico. Putin's assurances about the safety of oil extraction in the Arctic, however, are not very credible. Greenpeace for quite some time has been alerting about the state of the infrastructure of Russian oil platforms and the possibility of an ecological disaster in the Arctic.

The Rationale of the Canadian–Danish Coalition

In view of the chances of profits from the extraction of energy resources in the Arctic, the biggest and most emphasized problem is still the unresolved dispute over the Lomonosov Ridge. Admittedly, there exists a nuanced difference between "vying for ownership" and "being granted the rights to exploit the resources" of the outer continental shelf. Nevertheless, it is commonly understood as being rightfully present there with all the accompanying privileges granted. Canada, Russia, and Denmark, and therefore also Greenland, claim it is their own. Each of them invested huge sums of money in research that could confirm that it is part of their shelf. Russia was the first to make the claims, and as early as 2001, it submitted a request to the Continental Shelf Commission, but its form of the request was challenged. Then the Russian Federation reapplied in 2002. This means that Russia was the first to give up cooperation in the Arctic – giving a signal to the other participants that it was choosing the conflict and the benefits of extracting raw materials in the Arctic. Interestingly, the other participants did not respond directly to Russia's gambit and never gave any indication that it was necessary to abandon the cooperation and bet on the conflict; however, it should be assessed that they started to prepare slowly for such a game.

The new maps drawn by Canadian scientists were created both in Ottawa and during the expedition in the Far North, where icebreakers and helicopters were used to make maps of the bottom of the sea at many kilometers deep and as accurate as possible. The research involved Russian, American, and Danish scientists – countries that are vitally interested in determining the seabed as a place where valuable resources can be exploited in the future. The help of an unmanned submarine, which analyzed the shape of the seabed under the ice, turned out to be extremely important. In fact, this can be treated as a peaceful attempt to resolve the dispute.

To prove their claims to the Lomonosov Ridge in 2012, a team of 20 Danish scientists set out from Svalbard on board the Swedish icebreaker *Oden*. Their task was to prove that 155 square kilometers of the bottom of the Arctic Ocean, including the North Pole, is an extension of the Greenlandic continental shelf and should, therefore, be incorporated into the Danish Realm. The aim of the scientific expedition was to collect bathymetric data on the Lomonosov Ridge and sediment layers in the Amundsen Basin in the eastern part of the North Pole. The gathered data were to allow Denmark to demand from the UN to grant it a fairly large area of the seabed stretching north of the current 200 nautical miles from Greenland to the North Pole. According to Danish scientists, they have evidence to connect Greenland to the Lomonosov Ridge. Thanks to this, the government and Danish diplomacy have the evidence to demand a 200-mile exclusive economic zone off the coast of the island and exploitation of raw materials on the ocean floor.

Russia wanted to seal its sovereignty in this region. Already in 2007, the expedition of the Russian ship *Akademik Fedorov* took on a special dimension due to the placement of the titanium flag on the bottom of the Arctic Ocean at the North Pole. At that time, the following headlines appeared in the international press: *The Fight for the Pole, the Cold War in the Arctic.* Russia stated that this was a symbolic gesture, similar to the placing of the flag on the moon by the Americans. As Janusz Symonides notes,

> *...there is such a concept as symbolic appropriation in international law. In the past, the discoverer often expressed the symbolic appropriation of a place by placing the flag, building a fortification or installation. But later on, the international community began to recognize the title to new or disputed areas of land occupied by the state only on the basis of effective occupation.*

Such steps were also taken by the Russian side in October 2014, which announced that it was able to fully control the entire coastal zone in the Arctic region, stretching over 6,200 kilometers. Evidence of this was the modernization of seven air bases abandoned in 1991, the construction of a radar network from Murmansk to Chukotka, and continual exercises of the Russian fleet, air and land forces in the Arctic.

Canada and Denmark waited until the last minute, delivering partial submissions to the Commission in 2013 and 2014. The Russian side responded in accordance with Nash's logic, i.e. in a negative way to the position of Denmark and Canada. On December 17, Sergey Donskoy announced that his Ministry of Natural Resources and the Environment will prepare appropriate documents to submit an application to the UN for recognition of Russian sovereignty over a significant area of the Arctic. In August 2015, Russia delivered revised submission to the Commission on the Limits of the Continental Shelf (CLCS) in respect of the Arctic Ocean. Canada also responded with a modified submission in May 2019 that proposed new outer limits. The Commission reviews the submission in order they received. Russia is first on the CLCS's list, with the first partial submission delivered in 2001, Canada and Denmark are at the end of the list, 76 and 84, respectively (UN, 2020). It means that it might take years for the submission to be reviewed. Additionally, Danish's, Canada's, and Russian's submission overlap. They all included the North Pole in their submissions. According to the United Nations Convention on the Law of the Sea (UNCLOS), the states with overlapping claims must negotiate separate delimitation agreements, so the delimitation process will be further prolonged.

If any country's application is accepted, it will probably take several decades. One cannot be sure of Russia's peaceful intentions and the fact that it will recognize the claims to this territory, be it Canadian or Danish.

According to the logic of Nash, we can predict that we will be observers of the next round of conflict.

The Rationale for a Coalition between Denmark and the United States

It would seem that the alliance indicated is of no international significance. These countries at first glance differ significantly both in terms of territorial, population, economic and military potential, and geographical distance from each other. However, their alliance may be of key importance in the balance of power, being an important factor blocking Russia's intentions toward the Arctic. Greenland will play a key role in this arrangement. Both Denmark and the United States are aware of this fact, as evidenced by the alliances concluded by the parties in the past and nowadays, reaching even the period of World War II (WWII).

When the German aggression against Denmark took place in 1940, the United States started to fear that Germany may demand colonial possessions of Denmark, including Greenland. Its geographical location, close to America, was, therefore, a direct threat to the United States at that time. Hence on April 10, 1940, Franklin D. Roosevelt issued a statement, after consultations with the Ambassador of Denmark, Henrik Kauffmann, and with the consent of Greenland's local authorities, that the United States included Greenland in its sphere of influence. As Peter Bjerregaard and T. Kue Young indicate, Greenland's contacts with the United States were a breakthrough in its isolation in the international arena. The agreement concluded between the parties gave the United States the right to establish military bases in Greenland, including the air base in Thule. It is worth noting that in the period of WWII, the island was definitely influenced by both the United States and Canada, which provided it with food in exchange for the supplies of cryolite. Moreover, this coalition allowed the United States to take control of the North Atlantic, which was of an utmost importance against the Axis countries. In turn, after the WWII and in the face of growing Cold War, the United States brought forth a proposal to buy Greenland for 100 million dollars in gold, but Denmark refused. The establishment of the North Atlantic Treaty Organization (NATO) in 1949 strengthened the alliance of the countries under investigation. Thanks to that, the Americans negotiated a new agreement with Denmark on April 27, 1951, as regards building and using the existing military bases in Greenland. Already then Greenland played a significant role in maintaining the Cold War order, giving the Americans an advantageous position in the conflict with Russia.

It is worth stressing that nowadays the cooperation of the discussed States may be of a synergic character, and the countries with such different potentials may complement each other, using their strengths in the international arena.

In this context, Greenland is also important. History has shown its strategic prominence, which also has its consequences today. This is evidenced by the agreement, which was contracted in 2002 between Denmark, Greenland, and the United States to expand the base in Thule. Thanks to it, the agreement of 1951 was renegotiated.

The second issue is the affiliation of partners with NATO. It is difficult to say nowadays that there is effective multilateral cooperation within the Arctic Eight or the Arctic Five. The key player in the Arctic is Russia. Trying to understand Russia's intentions in the Arctic and its strategy for action in Central and Eastern Europe, cooperation for security purposes is becoming important in this region. In May 2015, a group of Norwegian experts, led by Rolf Tamnes from the Norwegian Institute of Defense Research, prepared a report for the Norwegian Ministry of Defense indicating that the annexation of the Crimea to Russia, as well as a part of eastern Ukraine, has implications for the security situation. According to the report, the annexation of the Crimea provides evidence that Russia may try to take control and deny other countries access to the Kola Peninsula, as well as to the adjacent areas of northern Norway and the Barents Sea and Norwegian Sea. The report did not rule out tensions related to border demarcation on the Arctic shelf and the Svalbard archipelago.

When Ukraine wanted some definitive action from the United Europe, nothing happened. Although sanctions were imposed on Russia, they did not change this country's strategic policy. There was no significant change of direction in the policy of the largest European corporations toward Russia. Neither did it reevaluate the policies of European countries with interests in Russia. This is a sign that the only entity that will be important in the Arctic region today is not the EU, but the United States, with the combination of NATO. The examples presented above show relations between the States, which extend far beyond the Arctic. They allow us to understand Russia's intentions in the region – and this country's own interests beyond the common good. Russia, therefore, appears in the international arena as a two-track player. On the one hand, its involvement in various international organizations and institutions can and should be noted, but incidents with NGOs such as Greenpeace and WWF show its different face.

It may be important in the case of Russia's aggression against any NATO State that point 5 of the Treaty declares that

> *The Parties agree that an armed attack against one or more of them in Europe or North America shall be considered an attack against them all and consequently they agree that, if such an armed attack occurs, each of them, in exercise of the right of individual or collective self-defence recognised by Article 51 of the Charter of the United Nations, will assist the Party or Parties so attacked by taking forthwith, individually and in concert with the other Parties, such action as it deems necessary, including the use of armed force, to restore and maintain the security of the North Atlantic area.*

The annexation of Crimea showed how unpredictable Russia is. Hence, in response, the NATO countries are trying to show Russia their military potential. To this end, they are organizing a growing number of exercises on military cooperation in this region. In 2018, for example, the scenario assumed a violation of Norwegian sovereignty. The main part of the exercises of the air and sea forces began on October 25th, but they were preceded by a preparatory stage on the shores of Iceland and northern Norway. Behind the polar circle, among other things, the landing of American infantry under the cover of aircraft from the *USS Harry Truman* nuclear carrier was practiced. In total, about 50,000 troops, 150 planes and helicopters, and 70 warships from 31 countries participated, including neutral Sweden and nonalliance Finland – but close coalition partners of Denmark within the frame of the Nordic Council and Nordic Defence Cooperation (NORDEFCO).

CONCLUSIONS

The Earth's climate has changed throughout history. Just in the last 650,000 years, there have been seven cycles of glacial advance and retreat, with the abrupt end of the last ice age about 7,000 years ago marking the beginning of the modern climate era and of human civilization. Most of these climate changes are attributed to very small variations in Earth's orbit that change the amount of solar energy our planet receives.

Humanity has not always understood that climate changes may have been due to natural causes and there was nothing people could do about them. They attributed them to divine interventions, wrath of gods and tried to appease them, or the devil or aliens meddling in our earthly affairs. Mother Nature invariably took care of the climate and things always returned to "normal," though not necessarily at the same places and for the same people. We will not live long enough to see Nature reacting to the changes and taking care of the system. That is exactly why quite a few books on the Arctic have the word "imperative" in them. If we do not act now, we are heading toward a disaster, and knowingly at that.

It is very true that during the last 30 years, the Arctic region has undergone fundamental transformation primarily due to climate change. As Paula Kankaanpaa and Oran R. Young alleged, it transformed from a frozen wasteland into the "last global frontier" of economic opportunity (Kankaanpaa, & Young, 2012, pp. 379–397). Given the amount of works devoted to the subject of the Arctic, this book is naturally only one of the many, hopefully adding to the body of knowledge and presenting the problem from a somewhat different angle. The region has become famous because of the rapidly changing weather patterns. The Arctic has also served another purpose: it has brought to humanity the awareness of interconnectedness of everything that is occurring on our planet. The ambition of the authors was to provide a compendium of fundamental information and data on politics and development in the North American Arctic by not only presenting the facts but also providing political, legal, economic, and social frame of the contemporary processes.

To examine the frame of the processes, J. Nash's game theory was used, which allowed to analyze politics and development in the North American Arctic, due to climate change, from two perspectives: cooperation and competition.

Through the prism of John Nash's theory, major challenges and opportunities that lead to cooperation and rivalry in the region have been suggested. We have also identified the political, social, and economic distinctive features of the North American Arctic which include decentralized systems of governance that comprise Indigenous Population (especially Canadian and Danish polices are shaped by the native inhabitants of the North); scattered population, limited infrastructure; social and economic instability; high level of financial dependency of subnational governments (except for Alaska); common languages: English (even in Greenland); and the Inuit languages.

An interaction between the melting of the Arctic Ocean and the international competition to access the resources and shipping routes put the states in the North American Arctic under substantial pressure. This weight combined with the quest for democratic self-governments transformed the system of governance in the region. It is of utmost importance that the power allocated to subnational and indigenous governments has institutionalized them as an important player in the management of the North American Arctic. Until the 1960s, across this region, the subnational governments had a very limited authority, and the relations between the Indigenous peoples and the central governments were generally based on colonial principles. Due to decolonization, democratization, and decentralization processes, the Canadian territories were granted more responsibilities, Greenland became a county of Denmark, and Alaska gained the statehood. Then the devolution of land and resources and self-government agreements followed. Therefore, today subnational governments function autonomously according to the democratic principles. The Inuit and other Indigenous peoples in the North American Arctic have formed various forms of self-government. In the case of Greenland, the Inuit self-government practically took over the decision-making process, similarly in Nunavut. Alaska's North Slope and Arctic Slope have borough governments with a legislative assembly and planning commissions, similar to a county within American states. On the national level, the North American Arctic States adjusted its strategies and policies toward the Arctic based on national objectives, international developments, and climate pressure. Canadian Prime Minister Stephen Harper put forward a very ambitious plan to protect sovereignty in the North and strongly promoted it. "Arctic sovereignty" became a key phrase in the political discourse during Stephen Harper's decade and is still very much present today. The then Canadian government's actions left the impression that the Northern territory was at risk. Small disputes with Denmark and the United States did not pose any serious threats. On the contrary, the Canadian authorities put more strain on these relationships by renaming the Northwest Passage to the Canadian Passage or deciding to claim the North Pole. Justin Trudeau's policy is largely based on the predecessor's concept, but the Liberal government evoked the value of multilateralism to the Arctic. Denmark's Arctic policy had to take into account the arrangements of the new self-government in Greenland. Launched in 2011, the Denmark Strategy for the Arctic 2011–2020 was prepared with cooperation of Greenlandic self-government. The Strategy called for global cooperation on the Arctic, including focus on climate change, protection of nature and environment, endorsement of strict global maritime rules, and protection of Indigenous peoples' rights not only within the frame of the Arctic Council but also globally. It also underlined the value of bilateral cooperation with Canada and the United States. The US Arctic Strategy of 2013 was based on or rather informed by the guiding principles of safeguarding peace and stability, making decisions by using the best available information, pursuance of innovative arrangements, and consultations and coordination with Alaska Natives. The US Arctic is built on three fundamental lines of effort, all of which are to advance US interests: security, pursuance of responsible Arctic Region stewardship, and strengthening international cooperation. Nevertheless, the United States seems to have retained the old posture of being a "reluctant" Arctic country and – as Americans would have phrased it – putting Arctic issues on a back burner.

When we compare current Canadian, Danish, and American policies and strategies toward the Arctic, we can identify well-defined goals and objectives. All the governments acknowledge that climate change profoundly impacted the region, so they try to respond to it. All of them prioritized the issue of security, sustainable development, and international cooperation in the region, although Canada and Denmark added a well-being of indigenous population as a main concern. The international cooperation especially through the Arctic Council, which is the major forum of negotiations, complemented by the international law that applies to the Arctic Ocean, is important to all the parties.

Is it therefore possible to have a coalition of the North American Arctic States? Only one of the four options speaks of the balance of Nash. It is a situation in which entities cooperate with each other. This option is beneficial to all entities – but most difficult to achieve. In conclusion, the analysis of the prisoner's dilemma shows that conflict outweighs cooperation, although such a strategy is far less advantageous for the players. To achieve maximum winnings, a common strategy and cooperation between partners is needed.

Our findings show that North American Arctic is a living theater of cooperation and rivalry between subnational, national, and international actors within the frame of international and domestic law. They cooperate not only through the Arctic Council but also through several nongovernmental organizations (NGOs) such as IASC, Greenpeace, WWF, and ICC, mainly focusing on nature and environmental protection. There are disputes over the Hans Island, the Beaufort Sea boundary, and the status of the Northwest Passage, but the States try to balance their interest by rather testing and checking the opponents. The North American Arctic States highlight the value of regional cooperation but either within the whole Arctic region or bilaterally. It seems that the States do not pay attention to the geographical region and do not foresee it as a platform for joint activities. In John Nash's theory, the players more often decide on conflict than cooperation, while in the case of North American States, it seems that they try to keep the balance.

Therefore, considering everything, we could conclude the following:

(1) There is a noticeable division in the Arctic into actors who prioritize cooperation for the climate and on behalf of the Indigenous peoples and actors who declare an interest in natural resources and their extraction. This creates a division between those who want to cooperate and those who want to compete.

(2) There exist premises for cooperation between actors in the North American Arctic (especially Indigenous peoples and NGOs who can play an important role in this process).

(3) Environmental protection plays an important role in cooperation. The benefits of climate protection and conservation of natural resources in the Arctic are a premise for the cooperation of the North American Arctic States.

(4) It would seem that any sort of alliance between the North American States is hardly possible due to significant differences in terms of territorial, population, economic, and military potential and geographical distance from each other, i.e., a pronounced disparity, which is obvious. However, their cooperation may be of key importance in the equilibrium, counterbalancing Russia in the Arctic – always assuming that balance is good.

It is worth stressing that nowadays the cooperation of the discussed States may be of a synergic character and the countries with such different potentials may complement each other, using their strengths in the international arena.

This is a sign that the only entity that will be important in the Arctic region today is not the European Union, but the United States, with the combination of NATO. The examples presented above show relations between the States, which extend far beyond the Arctic. They allow us to understand Russia's intentions in the region – and this country's own interests beyond the common good. Russia therefore appears in the international arena as a two-track player. On the one hand, its involvement in various international organizations and institutions can and should be noted, but examples of NGOs such as Greenpeace and WWF show its different face.

However, it is difficult to indicate that countries are willing to or giving up their rights to the Arctic. This may bring conflicts between the surveyed entities. As has been shown, the North American Arctic States both cooperate and compete with each other. This happens alternately, and the parties often take international law into account in their actions. Therefore, it is difficult to confirm that the Nash theory works in these cases and subsequent rounds contribute to deepening conflicts. Thus, it can be predicted that relations in the region will continue changing and balancing between conflict and cooperation. The benefits of international cooperation, which are felt not only by the Arctic States but also by all the actors of international relations, may be important in this context.

It seems that all of the Arctic governments should take into considerations the statement of the former US ambassador to Canada David Jacobson that the real threats in the North do not come from other nations. They come from the two challenges that have existed since man first set foot in that region: hostile weather and vast distances. And the only way Canada and all other affected nations are going to deal with those challenges is by the way that Indigenous people have dealt with them from time immemorial: cooperation (Chase, 2014).

The authors are fully aware that what appeared true at the time of writing the book may prove much outdated tomorrow and definitely fall into the category of yesterday's new. In view of the above, the work focuses more on the historical processes and noted attitudes, rather than on predictions of the possible developments, always keeping in mind that the all-powerful Mother Nature will always have the last word. Given that, it does not release the mankind from the responsibility of trying to predict and prepare and attempt to answer the all-important question: should we rather adapt or ought we to mitigate or perhaps balance the two approaches? – keeping in mind Spinoza's observation:

> *One and the same thing can at the same time be good, bad, and indifferent, e.g., music is good to the melancholy, bad to those who mourn, and neither good nor bad to the deaf.*

REFERENCES

ACIA. (2004). Impacts of a warming Arctic: Highlights. Arctic Climate Impact Assessment. Retrieved from: https://www.amap.no/documents/doc/impacts-of-a-warming-arctic-highlights/792

ACIA. (2005). *Arctic climate impact assessment.* New York, NY: Cambridge University Press.

Act on Greenland Self-Government. (2009). Act no. 473 of 12 June 2009. Retrieved from aalakkersuisut.gl//~/media/Nanoq/Files/Attached%20Files/Engelske-tekster/Act%20on%20Greenland.pdf

Agreement on the Conservation of Polar Bears. (1973). Oslo. Retrieved from https://polarbearagreement.org/index.php/about-us/1973-agreement

Agukkasuk, T. (2012). Sivutmun – Moving forward: Beyond colonial schooling models for Alaska native peoples. Retrieved from https://alaskaindigenous.wordpress.com/page/5/

AIP. (2015). Self-government agreement-in-principle, Inuvialuit Regional Corporation. Retrieved from https://irc.inuvialuit.com/self-government/self-government-aip#:~:text=The%20AIP%20confirms%20that%20the,and%20services%20to%20the%20Inuvialuit

Alaska Statehood Act. (1958) Public Law 85-508, 72 Stat. 339, July 7, 1958.

Alcantara, C. H., Cameron, K., & Kennedy, S. (2012). Assessing devolution in the Canadian North: A case study of the Yukon Territory. *Political Science Faculty Publications*, 29. Retrieved from https://scholars.wlu.ca/poli_faculty/29

AMAP. (2020). Geographical coverage. Retrieved from https://www.amap.no/about/geographical-coverage

Anderson, J. (2016, November 9). The US quietly re-writes Arctic policy. *Global Risk Insights.* https://globalriskinsights.com/2016/11/23956/

Anon. (n.d.). About IASC. Retrieved from https://iasc.info/iasc/about-iasc

Arctic Climate Impact Assessment. (2004). *Impacts of a warming Arctic: Highlights*. Cambridge: Cambridge University Press. Retrieved from https://www.amap.no/documents/doc/impacts-of-a-warming-arctic-highlights/792

Arctic Council. (1996). Declaration on the establishment of the Arctic council. Joint communique of the governments of the Arctic countries on the establishment of the Arctic council. Retrieved from https://oaarchive.arcticcouncil.org/handle/11374/85

Arctic Council Scorecard. (2017). 2017 WWF, Arctic Council Conservation Scorecard. Retrieved from http://wwf-ap.org/apps/acscorecard/action/conservation/

Arctic Council Scorecard. (2019). WWF, Arctic Council Scorecard 2019. Retrieved from https://c402277.ssl.cf1.rackcdn.com/publications/1223/files/original/overview_A4_-_web.pdf?1556629466

Arctic Environmental Protection Strategy. Declaration on the Protection of Arctic Environment. (1991, June 14). Retrieved from http://library.arcticportal.org/1542/1/artic_environment.pdf

Arctic Governance in an Era of Transformative Change: Critical Questions, Governance Principles. (2010). Ways forward report of the Arctic Governance project. Retrieved from www.norden.org

Arctic Update. (2017). The US Arctic Research Commission. Retrieved from https://www.arctic.gov/arctic-update/

Arruda, G. M., & Krutkowski, S. (2017). Arctic governance, indigenous knowledge, science and technology in times of climate change: Self-realization, recognition, representativeness. *Journal of Enterprising Communities: People and Places in the Global Economy*, *11*(4), 514–528. [Adobe Digital Editions version]. doi:10.1108/JEC-08-2015-0041

Åtland, K. (2017). The building up of Russia's military potential in the Arctic region and possible elements of its deterrence. *Centre for Russian Studies*. Retrieved from http://r-studies.org/cms/index.php?action=news/view_details&news_id=43590&lang=eng

Axelrod, R., Brosnan, I. G., Leschine, T. M., & Miles, E. L. (2011). Cooperation or conflict in a changing Arctic? *Ocean Development & International Law*, 42(1–2), 173. doi:10.1080/00908320.2011.543032

Backhouse, C. (1999). *Colour coded. A legal history of racism in Canada 1900–1950*. Toronto, ON: Toronto University Press.

Baev, P. K. (2008). From west to south to north: Russia engages and challenges its neighbours. *International Journal: Canada's Journal of Global Policy Analysis*, *63*(2), 291–305. doi:10.1177/002070200806300205

Baev, P. K. (2018). Examining the execution of Russian military-security policies and programs in the Arctic. In V.-P. Tynkkynen, S. Tabata, D. Gritsenko, & M. Goto (Eds.), *Russia's far north: The contested energy frontier*. London; New York, NY: Routledge.

Baker, J. S., & Byers, M. (2012). Crossed lines: The curious case of the Beaufort Sea maritime boundary dispute. *Ocean Development & International Law*, *43*(1), 70–95. doi:10.1080/00908320.2012.647509

Balcerowski, S. (2015). Retrieved from https://historia.org.pl/2015/01/14/najgorszy-interes-w-historii-rosji-sprzedaz-alaski-w-1867-r/

Barker, D. (2009). Was the Alaska purchase a good deal? *University of Iowa*. Retrieved from http://www.news-releases.uiowa.edu/2009/november/David%20Barker-Alaska.pdf

Bender, J. (2016). Countries have been fighting over an uninhabited island by leaving each other bottles of alcohol for over 3 decades. *Business Insider*. Retrieved from https://www.businessinsider.com/canada-and-denmark-whiskey-war-over-hans-island-2016-1?IR=T

Berger, T. R. (1989). True north strong and free inquiry society. In T. R. Berger & D. Roche (Eds.), *The Arctic: Choices for peace and security: Proceedings of a public inquiry*. Vancouver; Seattle, WA: Gordon Soules Book Publishers.

Berger, T. (1991). *A long and terrible shadow: White values, native rights in the Americas, 1492–1992*. Vancouver: Douglas and McIntyre.

Bergh, K. (2012). The Arctic policies of Canada and the United States: Domestic motives and International context. *SIPRI Insights on Peace and Security*, *1*, 1–20. Retrieved from https://www.sipri.org/sites/default/files/files/insight/SIPRIInsight1201.pdf

Bird, K. J., Gautler, D. L., Houseknecht, D. W., Klett, T. R., Pitman, J. K., Moore, T. E., … Wandrey, C. J. (2008). Circum-Arctic resource appraisal: Estimates of undiscovered oil and gas north of the Arctic circle fact sheet 2008–3049. *USGS*. Retrieved from http://pubs.usgs.gov/fs/2008/3049/

Blair, B. (2017). US Arcticness 150 years after the Alaska purchase. *The Polar Connection*. Retrieved from http://polarconnection.org/alaska-purchase/

Bolkhovitinov, N. (1990). The Crimean war and the emergence of proposals for the sale of Russian America, 1853–1861. *Pacific Historical Review*, *59*(1), 15–49. JSTOR 3640094.

Bolstad, E. (2010). Changes in Congress leave Alaska with fewer earmarks. *Anchorage Daily News*. Retrieved from https://www.adn.com/alaska-news/article/changes-congress-leave-alaska-fewer-earmarks/2010/02/22/

Bone, R. M. (1992). *The geography of the Canadian North*. Toronto, ON: University of Oxford Press.

Boulègue, M. (2019). *Russia's military posture in the Arctic managing hard power in a 'low tension' environment*. Chatham House: The Royal Institute of International Affairs. Retrieved from https://www.chathamhouse.org/publication/russia-s-military-posture-arctic-managing-hard-power-low-tension-environment

Braga, A. (2020, August 7). We can either all win or we all lose. Retrieved from https://www.greenpeace.org/international/story/44577/we-win-or-we-lose-all/

Breum, M. (2020, February 20). Greenland's premier: "We must work towards independence". *High North News*. Retrieved from https://www.highnorthnews.com/en/greenlands-premier-we-must-work-towards-independence

Burgess, P. (2010). Foundations of the Russian Federation's state policy in the Arctic until 2020 and beyond. Retrieved from https://icr.arcticportal.org/index.php?option=com_content& view=article&id= 1791%3

Buzan, B., & Albert, M. (2017). On the subject matter of international relations. *Review of International Studies*, *43*(05), 898. doi:10.1017/S0260210517000262

Bye, H. G. (2020, September 17). "U.S. Security Policy change in the Arctic places Denmark under pressure", says foreign policy researcher. *High North News*. Retrieved from https://www.highnorthnews.com/en/us-security-policy-change-arctic-places-denmark-under-pressure-says-foreign-policy-researcher

Byers, M. (2013). *International law and the Arctic*. Cambridge: Cambridge University Press. doi:10.1017/CBO9781107337442

Byers, M. (2014). *Who owns the Arctic? Understanding sovereignty disputes in the North*. Madeira Park, BC: Harbour Publishing.

Chartier, D. (2018). Qu'est-ce que l'imaginaire du Nord? In É. Le Calvé-Ivičević & V. Polić (Eds.), *Beyond the 49th parallel: Many faces off the Canadian North*. Brno: Masaryk University.

Chase, S. (2014, January 17). Myth versus reality in Stephen Harper's northern strategy. *The Globe and Mail*. https://www.theglobeandmail.com/news/national/the-north/myth-versus-reality-in-stephen-harpers-northern-strategy/article16397458/

Christensen, M., Nilsson, A. E., & Wormbs, N. (Eds.). (2013). *Media and the politics of Arctic climate change*. London: Palgrave Macmillan UK.

Coates, K. (1985). *Canada's colonies: A history of the Yukon and Northwest territories*. Toronto, ON: James Lorimer & Company.

Czarny, R. M. (2013). *The imperative high north: Opportunities and challenges*. Prešov: Michal Vaško – Vydavatel'stvo.

Czarny, R. M. (2015). *The high north: Between geography and politics*. Cham: Springer International Publishing.

Dahl, J. (1986). Greenland: Political structure of self-government. *Arctic Anthropology*, *23*(1&2), 315–324.

Department of Defense. (2019). Report to congress. Department of defense Arctic strategy. Retrieved from https://media.defense.gov/2019/Jun/06/2002141657/-1/-1/1/2019-DOD-ARCTIC-STRATEGY.PDF

Department of Justice Canada. (1985). Arctic Waters Pollution Prevention Act R.S.C., c. A-12. Retrieved from https://laws-lois.justice.gc.ca/eng/acts/A-12/FullText.html

Dewar, B. (2009, July 1). Nunavut and the Nunavut land claims agreement-an unresolved relationship. *Policy Options*.

Dickason, O. P. (1989). Concepts of sovereignty at the time of first contacts. In L. C. Green & O. P. Dickason (Eds.), *The law of nations and the new world*. Edmonton: University of Alberta Press.

Dingman, E. M. (2014). Greenlandic independence: The dilemma of natural resource extraction. *Arctic Yearbook*. Retrieved from https://arcticyearbook.com/images/yearbook/2014/Scholarly_Papers/11.Dingman.pdf

Diubaldo, R. (1981). The absurd little mouse: When Eskimos became Indians. *Journal of Canadian Studies*, *6*, 34–41.

DMA. (2020). Navigation in Greenland. Retrieved from https://www.dma.dk/SikkerhedTilSoes/Arktis/SejladsGroenland/Sider/default.aspx

DNV GL. (2015). Marine environmental risk assessment – Greenland. Retrieved from https://fmn.dk/nyheder/Documents/arktis-analyse/bilag6-risikoanalyse-havmiljoe-environmental-risk-assessment-greenland.pdf

Encyclopedia Britannica. (2020). Arctic. Retrieved from https://www.britannica.com/place/Arctic

Enge, B. (2017, September 30). USA: Trump turns his back on the Arctic. *High North News*. Retrieved from https://www.highnorthnews.com/en/usa-trump-turns-his-back-arctic

Exner-Pirot, H. (2015, February 11). The road to the U.S. Arctic council Chairmanship. *World Policy*. Retrieved from http://worldpolicy.org/2015/02/11/the-road-to-the-u-s-arctic-council-chairmanship/

Exner-Pirot, H. (2019, September 12). Canada's new Arctic policy doesn't stick the landing. *RCI Eye on the Arctic*. Retrieved from https://www.rcinet.ca/eye-on-thearctic/2019/09/12/canada-arctic-northern-policy-trudeau-analysis/

Flake, L. E. (2015). Forecasting conflict in the Arctic: The historical context of Russia's security intentions. *The Journal of Slavic Military Studies*, *28*(1), 72. doi:10.1080/13518046.2015.998122

Fogarassy, T., & Wilson, C. (Firm). (2007). *The Alaska boundary dispute: History and International law*. Clark Wilson LLP.

Gabryś, M. (2016). "Gedania" i Przejście Północno-Zachodnie. In M. Gabryś & M. Paluszkiewicz-Misiaczek (Eds.), *Pani Anna w Kanadzie. Księga pamiątkowa dedykowana Pani Profesor Annie Reczyńskiej*. Kraków: Księgarnia Akademicka. Retrieved from https://www.researchgate.net/publication/311716671_Pani_Anna_w_Ka na dzie_Ksiega_pamiatkowa_dedykowana_Pani_Profesor_Annie_Reczynskiej

Gad, U. P. (2017). What kind of nation state will Greenland be? Securitization theory as a strategy for analyzing identity politics. *Politik*, *20*(3). doi:10.7146/politik.v20i3.97156

Gautier, D. L., Bird, K. J., Charpentier, R. R., Grantz, A., Houseknecht, D. W., Klett, T. R., … Wandrey, C. J. (2009). Assessment of undiscovered oil and gas in the Arctic. *Science*, *324*(5931), 1175–1179. doi:10.1126/science.1169467

Gessain, R. (1978). *Ammassalik ou La civilisation obligatoire*. Paris: Flammarion.

Gnaś, H. (2015). Spory terytorialne państw członkowskich NATO na obszarze Arktyki. Uwarunkowania, znaczenie oraz wpływ globalnego ocieplenia. *Przegląd Strategiczny*, *5*(8), 33. doi:10.14746/ps.2015.1.3

Gorbachev, M. (1987). *Mikhail Gorbachev's speech in Murmansk at the ceremonial meeting on the occasion of the presentation of the order of Lenin and the Gold star to the city of Murmansk*. Moscow: Novosti Press Agency. Retrieved from http://www.barentsinfo.fi/docs/Gorbachev_spe ech.pdf

Gosnell, R. (2018, July 03). What will the future hold for Arctic economics? Retrieved from http://cimsec.org/what-will-the-future-hold-for-arctic-economics/37034

Government of Alaska. (2000). DNR, fact sheet, title: Land ownership in Alaska. Retrieved from http://dnr.alaska.gov/mlw/factsht/land_fs/land_own.pdf

Government of Canada. (1977). 'Fishing zones of Canada (zones 6) order', Consolidated regulations of Canada. Retrieved from https://laws-lois.justice.gc.ca/eng/regulations/C.R.C.,_c._1549/page-1.html

Government of Canada. (2005). Canada and Denmark issue statement on Hans Island. Retrieved from https://www.canada.ca/en/news/archive/2005/09/canada-denmark-issue-statement-hans-island.html

Government of Canada. (2009). Canada's northern strategy: Our north, our heritage, our future. Retrieved from https://web.archive.org/web/20150320084935/http://www.northernstrategy.gc.ca/cns/cns-eng.asp

Government of Canada. (2019a). Canada and the circumpolar Arctic. Retrieved from https://www.international.gc.ca/world-monde/international_relations-relations_internationales/arctic-arctique/index.aspx?lang=eng

Government of Canada. (2019b). Canada's Arctic and northern policy framework. Retrieved from https://www.rcaanc-cirnac.gc.ca/eng/1560523306861/1560523330587

Government of Canada. (2020). Mineral and metals facts. Retrieved from https://www.nrcan.gc.ca/our-natural-resources/minerals-mining/minerals-metals-facts/20507

Government of Nunavut. (2017). Pan territorial vison for sustainable development. Retrieved from https://www.gov.nt.ca/sites/flagship/files/documents/pan- territorial_vision_for_sustainable_development_en.pdf

Government of Nunavut. (2019). Budget 2018–2019. Retrieved from https://www.gov.nu.ca/finance/information/budget-2018-19

Graham, B. (1997). *Canada and the circumpolar world: Meeting the challenges of cooperation into the twenty-first Century*. Report of the House of Commons Standing Committee on Foreign Affairs and International Trade. Canada: House of Commons.

Grant, S. D. (2010). *Polar imperative: A history of Arctic sovereignty in North America*. Vancouver, BC/Toronto, ON/Berkeley, CA: Douglas & McIntyre.

Gray, D. H. (1997). Canada's unresolved maritime boundaries. *IBRU Boundary and Security Bulletin*, *5*, 61–70.

Greenland Home Rule Act. (1978). Act No. 577 of 29 November 1978.

Greenland in Figures. (2018). Retrieved from http://www.stat.gl/publ/da/gf/2018/pdf/Greenland%20in%20Figures%202018.pdf

Greenpeace International. (n.d.). History and successes. Retrieved from https://wayback.archive-it.org/9650/20200212060529/http://p3-raw.greenpeace.org/international/en/about/history/

Greenpeace International. (2013, November 13). Save the Arctic, greenpeace international. Retrieved from https://www.youtube.com/watch?v=uIz2PgzJ9YM

Greenpeace International. (2014a, July 8). LEGO: Everything is NOT awesome. Retrieved from https://www.youtube.com/watch?v =qhbliUq0_r4

Greenpeace International. (2014b, July 19). Arctic sanctuary. *Arctic Sanctuary*. Retrieved from https://www.greenpeace.org/international/publication/6905/arctic-sanctuary/

Greer, D. (2019, September 18). Arctic framework same old, same old. *Kivalliq News*. Retrieved from https://nunavutnews.com/nunavut-news/arctic-framework-same-old-same-old/

Griffiths, F. (2009). Canadian Arctic sovereignty: Time to take yes for an answer on the northwest passage. In F. Abele, T. J. Courchene, F. L. Seidle, & F. St-Hilaire (Eds.) Northern exposure: Peoples, powers and prospects for Canada's North (Vol. 4). Montreal: Institute for Research on Public Policy.

Groves, S. (2012, May 14). U.S. Accession to U.N. Convention on the Law of the Sea unnecessary to develop oil and gas resources. *The Heritage Foundation*. Retrieved from https://www.heritage.org/report/us-accession-un-convention-the-law-the-sea-unnecessary-develop-oil-and-gas-resources

Grydehøj (2016). Navigating the binaries of Island independence and dependence in Greenland: Decolonisation, political culture, and strategic services. *Political Geography*, *55*, 102. doi:10.1016/j.polgeo.2016.09.001

Guo, R. (2006). *Territorial disputes and resource management: A global handbook*. New York: Nova Science.

Hall, W. E. (1894). *Treatise on international law*. Oxford: Clarendon Press, and A Treatise on the foreign Powers and Jurisdiction of the British Crown, Oxford: Clarendon Press, 1894.

Hassol, S. J. (2004). *Impacts of a warming Arctic (executive summary)*. Cambridge: Cambridge University Press.

Haycox, S. (1990). Truth and expectation: Myth in Alaska history. *Northern Review*, 6, 65.

Head, I. L. (1963). Canadian claims to territorial sovereignty in the Arctic regions. *McGill Law Journal*, *9*(3), 200–226.

Heininen, L. (2019). Calotte Academy. Retrieved from https://iasc.info/working-groups/social-human/activities#workshop-materials-2

Hensley, W. L. I. (2017). Russia gave up Alaska. Why? *Alaska Dispatch News*. Retrieved from https://www.adn.com/opinions/2017/04/01/russia-gave-up-alaska-why/

van Herk, A. (2018). The North within. In É. Le Calvé-Ivičević & V. Polić (Eds.), *Beyond the 49 parallel: Many faces off the Canadian North* (pp. 53–66). Brno: Masaryk University.

Higginbotham, J., & Spence, J. (2018). *The North American Arctic: Energizing regional collaboration and governance*. Special report, by the Centre for International Governance Innovation. Retrieved from https://www.cigionline.org/sites/default/files/documents/NAA%20Special%20ReportWEB.pdf

Hirose, Y. (2018). International co-operation in the Arctic region: The search and rescue and the Barents co-operation. *Eurasian Journal of Social Sciences*, *6*(4), 37. doi:10.15604/ejss.2018.06.04.003

How Climate Made History. (2015). *History, nature documentary hosted by Jan Andres*. Special Broadcasting Service Corporation.

Huebert, R. (1998). New directions in circumpolar cooperation: Canada, the arctic environmental protection strategy, and the Arctic council. *Canadian Foreign Policy Journal*, *5*(2), 37. doi:10.1080/11926422.1998.9673131

Huebert, R. (2009). United States Arctic policy: The reluctant Arctic power. *University of Calgary, The School of Public Policy – University of Calgary Publications Series*, *2*(2). Retrieved from https://ssrn.com/abstract=3053702

Hunt, E. (2017, June 02). Where next for the US and the Paris deal? *The Guardian*. Retrieved from https://www.theguardian.com/environment/live/2017/jun/01/donald-trump-paris-climate-agreement-live-news

Hwang, P. (2019, March 06). Northern territories have highest rate of alcohol and drug related hospitalization: Study. *CBC News*. Retrieved from https://www.cbc.ca/news/canada/north/substance-abuse-hospital-stays-north-1.5158918#:~:text=Canada's%20three%20northern%20territories%20have,for%20Health%20Information%20(CIHI)

IASC. (2020). The IASC handbook of procedures & guidelines. Retrieved from http://iasc.info/images/about/organization/IASC_Handbook_20200416.pdf

ICC. (2010). Nuuk 2010 declaration. Retrieved from http://library.arcticportal.org/1896/1/Nuuk-Declaration.pdf

ICC. (2014). Kitigaaryuit declaration. Retrieved from https://www.inuitcircumpolar.com/icc-kitigaaryuit-declaration/

ICC. (2018). Utqiaġvik declaration 2018. Retrieved from https://www.arctictoday.com/wp-content/uploads/2018/07/2018-Utigavik-Declaration.pdf

ICC. (n.d.). About ICC. Uniting Inuit since 1977. Retrieved from https://www.inuitcircumpolar.com/about-icc/

International Arctic Science Committee. About IASC. (n.d.). Retrieved from https://iasc.info/iasc/about-iasc

IPCC. (2001). *Climate change 2001: Synthesis report. Contribution of working groups I, II, and III to the third assessment report of the Intergovernmental panel on climate change*. London: Cambridge University Press.

IPCC. (2014a). Intergovernmental panel on climate change, summary for the policy makers, climate change: Impacts, adaptation, and vulnerability. Retrieved from http://www.ipcc.ch/pdf/assessment-report/ar5/wg2/ar5_wgII_spm_en.pdf

IPCC. (2014b). Assessment report, polar regions. Retrieved from http://www.ipcc.ch/pdf/assessment-report/ar5/wg2/WGIIAR5-Chap28_FINAL.pdf

IPCC. (2014c). *Climate change 2014: Synthesis report. Contribution of working groups I, II and III to the fifth assessment report of the Intergovernmental panel on climate change*. London: Cambridge University Press.

Jacobsen, M. (2016). Denmark's strategic interests in the Arctic: It's the Greenlandic connection, stupid! *The Arctic Institute*. Retrieved from https://www.thearcticinstitute.org/denmark-interests-arctic-greenland-connection/

Jerome, K. J. (2018). *Thoughts of an idle fellow* (1st ed.). Paderborn: Outlook Verlag GmbH.

Jonson, L. (1990). Jonson, L. Soviet policy towards Sweden and the region of Northern Europe under Gorbachev. Cooperation and Conflict. *Cooperation and Conflict*, *25*(1), 1. doi:10.1177/001083679002500101

Kankaanpaa, P., & Young, R. (2012). The effectiveness of the Arctic council. *Polar Research*, *31*(1), 379–397. doi:10.3402/polar.v31i0.17176

Keil, K. (2014). The Arctic: A new region of conflict? The case of oil and gas. *Cooperation and Conflict*, *49*(2), 162–190. doi:10.1177/0010836713482555

Kikkert, P., & Lackenbauer, P. W. (2019). Canada's Arctic and Northern policy framework: A roadmap for the future? *Arctic Yearbook*. Retrieved from arcticyearbook.com/arctic-yearbook/2019/2019-briefing-notes/332-canada-s-arctic-and-northern-policy-framework-a-roadmap-for-the-future

Kingdom of Denmark. (2011). *Kingdom of Denmark strategy for the Arctic 2011–2020*. Retrieved from http://library.arcticportal.org/1890/1/DENMARK.pdf

Klein, G. (2018). Reaching Arctic mines by sea. *Resource clip*. Retrieved from resourceclips.com/2018/09/10/reaching-arctic-mines-by-sea/

Knapp, G. (2012). *An introduction to the economy of Alaska*. University of Alaska Anchorage. https://iseralaska.org/static/legacy_publication_links/presentations/2012_02-Introduction_to_Economy_of_Alaska.pdf

Koivurova, T., Molenaar, E., & Vanderzwaag, D. L. (2009). Canada, the European Union, and Arctic Ocean Governance: A tangled and shifting seascape and future directions. *Journal of Transnational Law & Policy*, *18*(2), 247–287.

Lackenbauer, P. W. W. (2013). *The Canadian rangers. A living history*. Vancouver, BC: UBC Press.

Lackenbauer, P. W. W., & Kikkert, P. (2020). Measuring the success of the Canadian rangers, a report. *North American and Arctic Defence and Security Network c/o School for the Study of Canada Trent University*. Retrieved from https://www.naadsn.ca/wp-content/uploads/2020/10/Rangers-Success-Metrics-Lackenbauer-Kikkert-upload.pdf

Lalonde, S., & Lasserre, F. (2013). The position of the United States on the Northwest passage: Is the fear of creating a precedent warranted? *Ocean Development & International Law*, *44*(1), 28. doi:10.1080/00908320.2012.726832

Lorenz, E. N. (1972). Predictability: Does the flap of a Butterfly's wings in Brazil set off a Tornado in Texas? 1972 conference of the American association for the advancement of science. Washington, DC.

Łuszczuk, M. (2015). *Ewolucja ról międzynarodowych w Arktyce*. Lublin: Wydawnictwo Uniwersytetu Marii Curie-Skłodowskiej.

Makowski, A., & Rossa, D. (2011). Współpraca akwenowa w regionie Arktyki i Antarktyki. In D. Rossa (Ed.), *Współpraca akwenowa państw we współczesnych stosunkach międzynarodowych* Poznań: Instytut Naukowo-Wydawniczy MAIUSCULA.

Masson-Delmotte, V., Zhai, P., Pörtner, H.-O., Roberts, D., Skea, J., & Shukla, P. R. (2019). Global warming of 1.5°C. *Intergovernmental Panel on Climate Change*. Retrieved from https://www.ipcc.ch/sr15/download/

McFarland, K. M., & Lide, V. (2018). The Arctic is melting. Here's why cooperation and diplomacy get so complicated. *The Washington Post*.

McGwin, K. (2016). Solid majority favours Greenland independence. *High North News*. Retrieved from http://www.highnorthnews.com/solid-majority-favours-greenland-independence

McNeil, K. (2007). The jurisdiction of inherent right aboriginal governments. *Osgoode Hall Law School of York University*. All Papers. 261. Retrieved from https://digitalcommons.osgoode.yorku.ca/all_papers/261

Medby, I. A. (2017). UCL department of geography, "preface – 'arcticness and change". Preface to Ilan Kelman, I., Arcticness: Power and voice from the North. UCL Press.

MFAD. (2016). Greenland and Denmark present claims relating to the continental shelf to the United Nations in New York. Retrieved from https://um.dk/en/news/newsdisplaypage/?newsid=3a2bd941-d477-4df9-8ad7-ef6f15ae9ec8

Minik, R. (2014). To the benefit of Greenland. *University of Copenhagen & Ilisimatusarfik (University of Greenland)*. Retrieved from greenlandperspective.ku.dk/this_is_greenland_perspective/background/report-papers/To_the_benefit_of_Greenland.pdf

Ministry of Foreign Affairs. (1993). The Nuuk declaration on environment and development in the Arctic: Report. Retrieved from https://books.google.pl/books?id=uRXqtgAACAAJ

Murray, K. (2014). The Arctic council: A brief history, its partnership with Indigenous groups of the Arctic, and its role in uncovering and addressing health issues in the Far North. *Journal of the Canadian Health Libraries Association*, *35*(01), 27. doi:10.5596/c14-006

Myrdal, G. (1957). *Economic theory and underdeveloped regions*. London: Gerald Duckworth & Co.

Nash, J. F. (1950). Equilibrium points in n-person games. *Mathematics: G. Polya*, *36*(1), 48–49.

Navy Task Force Climate Change. (2014). *U.S. Navy Arctic roadmap 2014–2030*. Washington, DC: Office of the Chief of Naval Operations. Retrieved from http://www.navy.mil/docs/USN_arctic_roadmap.pdf

Northwest Territories Devolution Act. (2014). S.C. 2014, c. 2. Retrieved from https://laws-lois.justice.gc.ca/eng/annualstatutes/2014_2/FullText.html

Northwest Territories Strategy. (2020). Northwest territories economic opportunities strategy. Retrieved from https://www.iti.gov.nt.ca/sites/iti/files/0004-704_econ_opp_strat_-_low-res.pdf

O'Brien, K. L., & Leichenko, R. M. (2000). Double exposure: Assessing the impacts of climate change within the context of economic globalization. *Global Environmental Change*, *10*(3), 221–232. doi:10.1016/S0959-3780(00)00021-2

Olesen, M. R. (2017). *Arctic rivalries: Friendly competition or dangerous conflict?* Copenhagen: Danish Institute for International Studies. Retrieved from https://pure.diis.dk/ws/files/1141975/DIIS_WP_2017_6.pdf

Oppenheim, L. F. L. (1905–06). *International law: A treatise*. Oxford: Clarendon Press.

O'Rourke, R. (2012/2018). Coordinator specialist in naval affairs, changes in the Arctic: Background and issues for Congress. *Congressional Research Service*. 7-5700. Retrieved from www.crs.gov.R41153

Osica, O. (2010). Daleka Północ jako nowy obszar współpracy i rywalizacji. *Nowa Europa. Przegląd Natoliński, special issue*, *1*(4), 5–54.

Osiński, P. (2016). Najgorsza transakcja w historii. Jak Rosjanie sprzedali Alaskę. Retrieved from http://superbiz.se.pl/wiadomosci-biz/najgorsza-transakcja-w-historii-jak-rosjanie-sprzedali-alaske_862611.html

Østhagen, A. (2020). International cooperation as an Arctic solution? In A. Østhagen (Ed.), *Coast guards and ocean politics in the Arctic*. Singapore: Springer. doi:10.1007/978-981-15-0754-0_6. [Adobe Digital Editions version].

OutRight. (2017). UN program. Retrieved from https://www.outrightinternational.org/content/what-does-sovereignty-united-nations-really-mean-today

Palosaari, T. L. (2012). The Amazing race: On resources, conflict, and cooperation in the Arctic. In sustainable development in the Arctic region through peace and stability. *Nordia Geographical Publications*, *40*(4), 13–30. Retrieved from http://urn.fi/URN:NBN:fi:uta-201302011022

Parry, M. L., Canziani, O. F., Palutikof, J. P., Hanson, C. E., & van der Linden, P. (2007). *Intergovernmental panel on climate change. Climate change 2007: Impacts, adaptation and vulnerability*. Geneva: IPCC Secretariat.

PCIJ. (1933). Legal Status of Eastern Greenland (Den. v. Nor.), 1933 P.C.I.J. (ser. A/B) No. 53 (Apr. 5).

Pearson, L. (1946). Canada looks "down north". *Foreign Affairs*, *24*(4), 638–647. doi:10.2307/20029999

Pharand, D. (2019). The case for an Arctic region council and a treaty proposal. *Revue Générale de Droit*, *23*(2), 163–195. doi:10.7202/1057469ar

Powell, M. (2010). How Alaska became a federal aid magnet. *New York Times*. Retrieved from https://economix.blogs.nytimes.com/2010/08/18/how-alaska-became-a-federal-aid-magnet/?mcubz=0

Quote Investigator. (2010). Everybody talks about the weather, but nobody does anything about it. Retrieved from https://quoteinvestigator.com/2010/04/23/everybody-talks-about-the-weather/Quote Investigator. Accessed on February 13, 2012.

Raspotnik, A., & Uljua, R. (2018, January 09). Make Alaska and its petroleum great again? Business and economics, commentary, United States. Retrieved from https://www.thearcticinstitute.org/make-alaska-petroleum-great/

Reflections on the state of select scientific fields in Poland. (2010). Integrative and integrative expert teams of the Polish Academy of Sciences (PAN), Expert opinion No. 21.

Renner, G. T. (1942). Human geography in the air age. *The Macmillan Company*. Retrieved from https://archive.org/details/humangeographyin00rennrich/page/n5/mode/2up

Riddell-Dixon, E. (2017). *Canada sovereignty, and the extended continental shelf*. Toronto, ON: Dundurn.

Roach, J. A., Smith, R. W., & Roach, J. A. (2012). *Excessive maritime claims* (3rd ed.). Leiden, Boston: Martinus Nijhoff Publishers.

Rosenberg, E., Titley, D., & Wiker, A. (2014, December). Arctic 2015 and beyond, a strategy for U.S. Leadership in the high north, center for a New American Security. *Policy Brief*. Retrieved from cnas.org

Roston, E. (2018). The World has discovered a $1 Trillion Ocean. Retrieved from https://www.guggenheimpartners.com/GP/media/pdf/Bloomberg_The-World-Has-Discovered-a-1-Trillion-Ocean.pdf

S-G UN. (1999). Secretary general of the United Nations, two concepts of sovereignty. Retrieved from https://www.un.org/sg/en/content/sg/articles/1999-09-18/two-concepts-sovereignty

Scrivener, D. (1999). Arctic environmental cooperation in transition. *Polar Record*, *35*(192), 51–58. doi:10.1017/S0032247400026334

Shadian, J. M. (2018). The emerging economy of the North America Arctic. *Arctic, 360*. Retrieved from https://www.ourcommons.ca/Content/Committee/421/FAAE/Brief/BR10185960/br-external/ShadianJessica-e.pdf

Shaw, M. N. (2003). *International law*. Cambridge: University of Cambridge Press.

Slattery, B. (1983). The constitutional guarantee of aboriginal and treaty rights. *Queen's Law Journal*, *8*(1 & 2), 232. 1982–1983.

Smolińska, A. (2016). Kupno Alaski przez USA. *Najgorsza inwesycja w historii Ameryki*. Retrieved from http://www.ticker.pl/2016/08/10/kupno-alaski-przez-usa-najgorsza-inwesycja-w-historii-ameryki/

Spandau, A. (1968). Do the general laws of economics apply to underdeveloped countries? Gunnar Myrdal's "economic theory and underdeveloped regions" in retrospect. *The South African Journal of Economics*, *36*(3), 228–235. doi:10.1111/j.1813-6982.1968.tb02297.x

Spinoza, B. (1677). The ethics, part IV, preface. *The Project Gutenberg E-Bok*. Retrieved from https://www.gutenberg.org/files/3800/3800-h/3800-h.htm#chap04

Statista. (2020). Population projection for Canada in 2019 and 2043. Retrieved from https://www.statista.com/statistics/481509/canada-population-projection-by-province/

Statistics Canada (2019). Population estimates on July 1st 2019, Canada, Province or territory. Statistics Canada Data Table: 17-10-0005-01.

Statistics Canada. (2020). GDP at basic prices, by sector and industry, provincial and territorial. Statistics Canada Data Table: 36-10-0222-01.

Stevenson, C. (2007). Hans off!: The struggle for Hans Island and the potential ramifications for International border dispute resolution. *Boston College International & Comparative Law Review*, *30*(1), 263–274.

Stokke, O. S., & Hønneland, G. (Eds.). (2007). *International cooperation and arctic governance: Regime effectiveness and northern region building*. London: Routledge.

Strawiński, P. (2017). Zakup alaski przez USA. Historia sprzedaży alaski przez Rosjan. Retrieved from http://biznes.onet.pl/wiadomosci/swiat/zakup-alaski-przez-usa-historia-sprzedazy-alaski-przez-rosjan/tlv2x5

Sykulski, L. (2009). *Geopolityka. Słownik terminologiczny*. Warszawa: Wydawnictwo Naukowe PWN.

Territorial Outlook Economic Forecast. (2019). Retrieved from http://www.miningnorth.com/chamber-news/102025

The Arctic Council. (2020). The Arctic council. About. Retrieved from https://arctic-council.org/en/about/

The Arctic Institute. (2013). Center for circumpolar security studies. Retrieved from http://www.thearcticinstitute.org/

The Causes and Consequences of a Rapidly Changing Arctic. (2018). 2018: The causes and consequences of a rapidly changing Arctic. Retrieved from https://www.globalchange.gov/sites/globalchange/files/ Taylor_NOAAwebinar_final%20%28small%29.pdf

The United Nations. (1973). Agreement between the Government of the Kingdom of Denmark and the Government of Canada relating to the delimitation of the continental shelf between Greenland and Canada. Retrieved from http://www.un.org/Depts/los/LEGISLATIONANDTREATIES/PDFFILES/TREATIES/DNK-CAN1973CS.PDF

The White House. (2009). Presidential directives nspd-66/hspd-25, "subject: The Arctic region". Retrieved from www.fas.org.irp.offsdocz/nspd/nspd-66.htm

The White House. (2013, May 10). The National strategy for the Arctic region, Washington. Retrieved from https://obamawhitehouse.archives.gov/sites/default/files/docs/nat_arctic_strategy.pdf

Theutenberg, B. J. (1984, December). Mare Clausum et Mare Liberum. *Arctic*, *37*(4), 481.

Tomala, M. (2015). The role of greenpeace in ensuring environmental safety in the Arctic. *Studia i Materiały. Miscellanea Oeconomicae*, 2(4), 49–60.

Tomala, M. (2016). *Grenlanddia w systemie stosunków międzynarodowych w świetle teorii Immanuela Wallersteina systemów-światów*. Kielce: Jan Kochanowski University.

Ugarriza, J. E. (2009). Ideologies and conflict in the post-Cold War. *International Journal of Conflict Management*, *20*(1), 82–104.

UN. (2020). Submissions, through the Secretary-General of the United Nations, to the Commission on the limits of the continental shelf, pursuant to article 76, paragraph 8, of the United Nations convention on the Law of the Sea of 10 December 1982. Retrieved from https://www.un.org/Depts/los/clcs_new/commission_submissions.htm

UNCLOS. (1982). Part VI. Retrieved from https://www.un.org/depts/los/convention_agreements/texts/unclos/closindx.htm

United Nations. General Assembly. (1992). Report of the United Nations Conference on environment and development. Retrieved from https://www.un.org/en/development/desa/population/migration/general assembly/docs/globalcompact/A_CONF.151_26_Vol.I_Declaration.pdf

U.S. Coast Guard. (2013). United States Coast Guard Arctic strategy. Retrieved from http://polarconnection.org/united-states-coast-guard-arctic-strategy-may-2013/

USGS. (2007). Assessment of undiscovered oil and gas resources of the East Greenland Rift Basins. Retrieved from https://pubs.usgs.gov/fs/2007/3077/pdf/FS07-3077_508.pdf

U.S. House. (2015). An act making appropriations for the Department of Defense for the fiscal year ending and for other purposes. 30.09. HR 4870. 113th Cong., 2nd sess. Retrieved from http://www.gpo.gov/fdsys/pkg/BILLS-113hr4870rs/pdf/BILLS-113hr4870rs.pdf

U.S. Senate. (2015). Coast Guard authorization act for fiscal years 2015 and 2016. S 2444. 113th Cong., 2nd sess. Retrieved from http://www.gpo.gov/fdsys/pkg/BILLS-113s2444is/pdf/BILLS-113s2444is.pdf

Van Rouveroy van Nieuwaal, M. (2014). Black ice: The story of the Greenpeace Arctic 30. Retrieved from https://www.youtube.com/embed/ZnqN6AaFE_I

Vaughan, A. (2014). Lego ends Shell partnership following Greenpeace campaign. Retrieved from https://www.theguardian.com/environment/2014/oct/09/lego-ends-shell-partnership-following-greenpeace-campaign

Wajzer, M. (2018). Odnajdywanie prostoty w złożoności. Wybrane zastosowania gier 2×2 w badaniach zjawisk politycznych. *Przegląd Politologiczny*, 2, 61–70. doi:10.14746/pp.2018.23.2.4

Walsh, J. E., Fetterer, F., Stewart, J. J., & Chapman, W. L. (2017). A database for depicting Arctic sea ice variations back to 1850. *Geographical Review, 107*(1), 89–107. doi:10.1111/j.1931-0846.2016.12195.x

Ward Ford, M. J., & Rude, R. (1999). ANCSA: Sovereignty and a just settlement of land claims or an act of deception. *Touro Law Review, 15*(2), Article 10.

Watkiss, P., Downing, T., Handley, C., & Butterfield, R. (2005). The impacts and costs of climate change. *European Commission DG Environment.*

Weber, B. (2018, February 01). Development in Canadian Arctic lags behind other northern regions. *Globe and Mail.* https://www.theglobeandmail.com/news/national/development-in-canadian-arctic-lags-behind-other-northern-regions-think-tank/article37814193/

Węsławski, J. M., Stempniewicz, L., & Kwaśniewski, S. (2007). Konsekwencje zmiany klimatu dla morskiego ekosystemu Arktyki Europejskiej. In A. Styszyńska & A. A. Marsz (Eds.), *Zmiany klimatyczne w Arktyce i Antarktyce w ostatnim pięćdziesięcioleciu XX wieku i ich implikacje środowiskowe.* Gdynia: Wydawnictwo Uczelniane Akademii Morskiej.

Wilson Rowe, E. (2013). A dangerous space? Unpacking state and media discourses on the Arctic. *Polar Geography*, *36*(3), 232–244. doi:10.1080/1088937X.2012.724461

World Population Review. (2020). Alaska population 2020. Retrieved from https://worldpopulationreview.com/states/alaska-population

Woś, R. (2012). Kupno Alaska przez USA. Najgorszy interes w historii Ameryki. Retrieved from http://forsal.pl/artykuly/630947,kupno-alaski-przez-usa-najgorszy-interes-w-historii-ameryki.html

Wrońska, I. (2012). Canadian Arctic policy: Determinants and trends. In R. M. Czarny, R. Kubicki, A. Janowska, & R. S. Czarny (Eds.), *The Northern spaces: Contemporary issues* (pp. 99–116). Warszawa-Kielce: Scandinavium.

Wrońska, I. (2018). Decade of Stephen Harper's norther policy: Achievements and failures. In É. Le Calvé-Ivičević & V. Polić (Eds.), *Beyond the 49th parallel: Many faces off the Canadian North* (pp. 253–264). Brno: Masaryk University.

WWF. (2017). WWF Arctic council conservation Scorecard. *Assessment Report.* Retrieved from http://wwf-ap.org/apps/site/templates/downloads/wwf-arctic-council-conservation-scorecard-WEB.pdf

WWF. (n.d.). Global Arctic programme a global response to a global challenge. Retrieved from https://d2ouvy59p0dg6k.cloudfront.net/downloads/global_arctic_programme_factsheet_2012_01_letter_english.pdf

Yong, O. R., & Einarsson, N. (2004). *Arctic human development report.* Akureyri: Stefansson Arctic Institute.

Young, O. R. (1992). *Arctic politics: Conflict and cooperation in the circumpolar north.* Hanover: Dartmouth College Press.

Young, O. R. (2005). Governing the Arctic: From cold war theater to Mosaic of cooperation. *Global Governance*, *11*(1), 9–15. JSTOR.

Young, O. R. (2009). Whither the Arctic? Conflict or cooperation in the circumpolar north. *Polar Record*, *45*(1), 73–82. doi:10.1017/S0032247408007791

Yukon Act. (2002) S.C. 2002, c. 7.

INDEX

Printed and bound by CPI Group (UK) Ltd, Croydon, CR0 4YY

13/09/2026

14961061-0005